LECTURES ON ANALYSIS
VOLUME I

MATHEMATICS LECTURE NOTE SERIES

J. Frank Adams	LECTURES ON LIE GROUPS
E. Artin and J. Tate	CLASS FIELD THEORY
Michael Atiyah	K-THEORY
Hyman Bass	ALGEBRAIC K-THEORY
Melvyn S. Berger Marion S. Berger	PERSPECTIVES IN NONLINEARITY
Armand Borel	LINEAR ALGEBRA GROUPS
Raoul Bott	LECTURES ON K (X)
Andrew Browder	INTRODUCTION TO FUNCTION ALGEBRAS
Gustave Choquet	LECTURES ON ANALYSIS I: INTEGRATION AND TOPOLOGICAL VECTOR SPACES II: REPRESENTATION THEORY III: INFINITE DIMENSIONAL MEASURES AND PROBLEM SOLUTIONS
Paul J. Cohen	SET THEORY AND THE CONTINUUM HYPOTHESIS
Eldon Dyer	COHOMOLOGY THEORIES
Walter Feit	CHARACTERS OF FINITE GROUPS
John Fogarty	INVARIANT THEORY
William Fulton	ALGEBRAIC CURVES
Marvin J. Greenberg	LECTURES ON ALGEBRAIC TOPOLOGY
Marvin J. Greenberg	LECTURES ON FORMS IN MANY VARIABLES
Robin Hartshorne	FOUNDATIONS OF PROJECTIVE GEOMETRY
J. F. P. Hudson	PIECEWISE LINEAR TOPOLOGY
Irving Kaplansky	RINGS OF OPERATORS
K. Kapp and H. Schneider	COMPLETELY O-SIMPLE SEMIGROUPS
Joseph B. Keller Stuart Antman	BIFURCATION THEORY AND NONLINEAR EIGENVALUE PROBLEMS

Serge Lang	ALGEBRAIC FUNCTIONS
Serge Lang	RAPPORT SUR LA COHOMOLOGIE DES GROUPES
Ottmar Loos	SYMMETRIC SPACES I: GENERAL THEORY II: COMPACT SPACES AND CLASSIFICATION
I. G. Macdonald	ALGEBRAIC GEOMETRY: INTRODUCTION TO SCHEMES
George W. Mackey	INDUCED REPRESENTATIONS OF GROUPS AND QUANTUM MECHANICS
Andrew Ogg	MODULAR FORMS AND DIRICHLET SERIES
Richard Palais	FOUNDATIONS OF GLOBAL NON-LINEAR ANALYSIS
William Parry	ENTROPY AND GENERATORS IN ERGODIC THEORY
D. S. Passman	PERMUTATION GROUPS
Walter Rudin	FUNCTION THEORY IN POLYDISCS
Jean-Pierre Serre	ABELIAN *l*-ADIC REPRESENTATIONS AND ELLIPTIC CURVES
Jean-Pierre Serre	ALGEBRES DE LIE SEMI-SIMPLE COMPLEXES
Jean-Pierre Serre	LIE ALGEBRAS AND LIE GROUPS
Shlomo Sternberg	CELESTIAL MECHANICS

A Note from the Publisher

This volume was printed directly from a typescript prepared by the editors, who take full responsibility for its content and appearance. The Publisher has not performed his usual functions of reviewing, editing, typesetting, and proofreading the material prior to publication.

The Publisher fully endorses this informal and quick method of publishing conference proceedings, and he wishes to thank the editors for preparing the material for publication.

LECTURES ON ANALYSIS

VOLUME I

Integration and Topological Vector Spaces

GUSTAVE CHOQUET

Université de Paris

Edited by

J. MARSDEN

University of California, Berkeley

T. LANCE and S. GELBART

Princeton University

W. A. BENJAMIN, INC.

New York 1969 Amsterdam

LECTURES ON ANALYSIS VOLUME I Integration and Topological Vector Spaces

Standard Book Numbers: 8053-6960-0 (Cloth)
8053-6961-9 (Paper)
Library of Congress Catalog Card Number 79-83802
Manufactured in the United States of America
12345 MR 2109

*The manuscript was put into production on March 14, 1969;
this volume was published on May 15, 1969*

**W. A. BENJAMIN, INC.
New York, New York 10016**

FOREWORD

Modern analysts use a great variety of basic tools, many of which are common to analysts in several branches. I have tried, during a one semester graduate course in Princeton, to present some of those tools that I found useful in Potential theory, Probability theory and Harmonic analysis; they include parts of functional analysis, integration theory and general topology, but I have avoided as much as possible, parts of these theories which, although elegant and interesting in themselves, do not have a large range of applications.

This book includes notes on this course,

also on several lectures which have been given by participants of the course to provide applications and illustrations.

These notes have been prepared by a very active team of auditors of the course; Dr. Jerry Marsdan, Tim Lance and Steve Gelbart who not only did the editorial work, but also suggested improvements. I thank them very cordially for their work.

My thanks also go to Anne Kenny, the librarian of Fine Hall who helped me daily to the wealth of the library, and to Joanne Beal, Rosa Kao Magliola and especially to Alta Zapf for their beautiful typing. I also thank Ollie Cullers and Glynis Marsden for preparing the final draft.

January 1968 Gustave Choquet.

PREFACE

This book is an expansion and revision of lecture notes from Professor Choquet's course in analysis given at Princeton during the fall term in 1967. Although details are now given and a few extra results and additional problems (with hints and solutions) have been added, the main content of the book is the same as that given in the lectures.

The book is intended for graduate students in analysis and for use as a reference text for research workers in the field. The book is also intended for a selected one term course at the graduate level; however to cover all the material in the book would probably require a year's course for the average student. A short course, for example, might consist of §§4, 5, 6, 10, 11, 12, 15, 18, 19, 21, 22, 25, 26, 27, 28, part of 31, 34 and 36.

There are no special prerequisites for the book, except a solid undergraduate background, possibly supplemented by some beginning graduate courses. Generally, if the reader can easily handle chapter one, he should be adequately prepared. This first chapter should help the reader judge what material he needs to read the book, to review material for him, or to indicate where he should begin reading.

The book cuts across many areas of analysis. There is an attempt made to clearly point out relationships between the different areas. The book stresses a wide range of techniques and proofs and gives many examples and applications of the use of the theory. To this end we have included solutions to problems which develop further techniques and applications.

In a book of this scope it is unlikely that all of the misprints and errors have been discovered

We have tried to make the text honest and reliable without being too pedantic.

The material of this book contains many time tested and important results, especially those important for applications. Needless to say, the book is not a complete text for all of analysis. For example little Fourier analysis, differential equations and distribution theory appears (although we do use these fields for comparison and examples).

The majority of the text develops the material fully, with complete proofs, examples and applications. There are a few things however which are supplementary or are sketched without proof to indicate areas allied to the text but outside its main stream. Such sections are starred.

For those wishing to learn a particular topic, the table of interdependence of sections should help. For example one can proceed directly to representation theory after reading (or knowing) §1-6, 10-12, §15, 19, 21, and 22. Again, chapter 9 does not depend on 8.

In the text itself, the sections are numbered consecutively and the definitions, propositions and theorems are numbered consecutively in each section. This method is particularly useful for reference. For example proposition 13.4 refers to the fourth item in §13. Also to help in isolating the important results we reserve the name "theorem" for the main results, "proposition" for secondary results and "lemma" for technical results needed for proofs. In referring to a book we use its title and author (the full reference is found in the bibliography) and for a technical paper we refer to, for example, Choquet [2].

As with the lectures, we have tried to preserve the spirit "of obtaining the important results with a minimum of delay by avoiding unnecessary technicalities and excursions and at the same time giving a complete and honest exposition."

The problems form an important part of the book. They are intended to develop techniques and a working knowledge of the text. There is a wide range of difficulty, some problems being trivial and others being quite challenging even to the best student. Hints and solutions are provided at the end of the book. When no hint is given, this means the problem is very easy. Many exercises are motivated by later needs and we do not hesitate to use these results. Some further problems without solutions are listed at the end of the book.

An important aspect of doing mathematics is to gain intuition and feel for a subject. It is useful to be able to guess whether a result is true or not and to have a plan in mind before logical reasoning ensues. In the text itself we have attempted to convey the intuition with heuristic discussions, especially in delicate areas such as the notion of a simplex. The problems are also designed this way. For example the reader might be asked to conjecture, to generalize, or to make a concept precise. Also we do not star the hard problems, for we feel that an important talent to be learned is the ability to distinguish between trivialities and the crux of a difficult problem. This is the situation faced in research.

As we mentioned, this book is intended for both the student and the expert. We offer some advice for both groups. First, for the student: this book contains a great number of subtleties which are mastered only after the expense of a good deal of practice and hard work. There are two main ways to achieve this mastery. First: work out the unproven statements given in the text. These are generally simple exercises. Second: try problems. One thing that should be avoided is to dismiss a problem with a quick survey of the solution. Many problems are, quite frankly, very hard and require a long session to solve well. The solutions we give are sketches and will be quite useless without considerable contact with the problem. On the other hand many problems are very easy. As has long been known, the only way to understand delicate and deep theorems is to struggle with a related problem, even if that struggle is unsuccessful.

The expert will quickly spot any new results or simplifications of old proofs so we will not make a list of these. Most of the new results have been announced or talked about, for example, in Choquet [17]. Some new results appear in the problems, although these are generally minor, but should not be overlooked. On the whole the book is intended as an exposition of recent rather than new results, mainly in the area of representation theory.

We would like to express our deep gratitude

to Professor Choquet for the warm, energetic and enthusiastic response he gave us. Thanks are due to the other members of the course for their participation and comments, and in particular to Neal Stoltzfus, Richard Parris, Barry Simon and Charles Fefferman.

May, 1968

Princeton, New Jersey

J. Marsden
T. Lance
S. Gelbart.

INTERDEPENDENCE OF SECTIONS

A section depends in a main way on all sections above it connected by a solid line. A dashed line indicates a minor dependence.

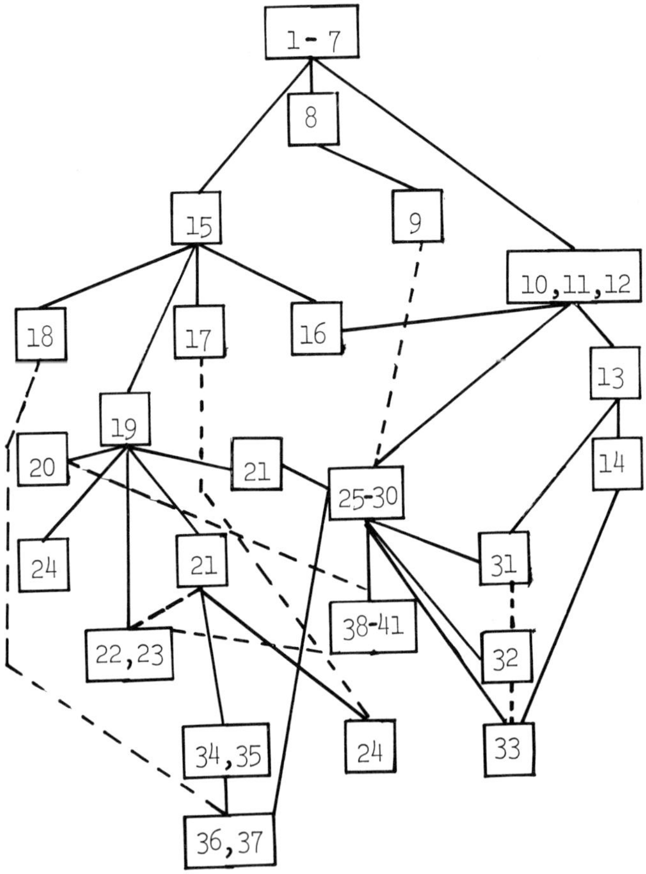

CONTENTS

VOLUME ONE

VOLUME TWO

VOLUME THREE

Important remark. The proof of the remarks following theorem 8.7 are included at the end of volume one as an appendix. As this was included after the rest of the manuscript was prepared, the subsequent apology in theorem 27.9 becomes unnecessary.

CHAPTER 1

PRELIMINARIES

This chapter is designed to be used as a reference and review, the selection of topics being dictated by our later needs. Proofs are omitted. The theorems are all standard, but are often enough forgotten or misunderstood to justify including them.

§1. SET THEORY

SETS AND CLASSES. The subject of logic and set theory is a complicated one which we cannot describe here. Rather, our starting point shall be the agreement that we work in the framework of naive set theory (see Halmos, Naive Set Theory).

Certain operations are permissible with sets, and others are not (in order to obtain a consistent set theory). We use the words set, family and collection synonymously. For example, if S is a set, then so is $\mathcal{P}(S)$, the collection of all subsets of S. In general if P is a "property" and S is a set then $\{x \in S: P(x) \text{ is true}\}$ is a set but merely $\{x: P(x) \text{ is true}\}$ is not a set; rather it is a less restrictive object, called a *class*. For example the collection of all groups is a class, not a set. The usual set theoretic operations may not be valid for classes, but we can speak of *members* of a class.

We are thus assuming that the reader is familiar with mappings and symbols $\emptyset$, $\in$, $\subset$, $\supset$, $\cap$, $\cup$, $\times$, $\setminus$, etc. If I is a set and

$\{X_\alpha : \alpha \in I\}$ is a family (set) of sets, there is a set S (by assumption) such that $X_\alpha \subset S$ for all α, and we define:

$$\cap\{X_\alpha\} = \cap\{X_\alpha : \alpha \in I\} = \{x \in S : x \in X_\alpha \text{ for all } \alpha \in I\}$$

and

$$\cup\{X_\alpha\} = \cup\{X_\alpha : \alpha \in I\} = \{x \in S : x \in X_\alpha \text{ for some } \alpha \in I\}.$$

Occasionally we shall write $\bigcap_{\alpha \in I} X_\alpha$ for $\cap\{X_\alpha : \alpha \in I\}$. These symbols are used only when dealing with sets and not classes.

For the family $\{X_\alpha : \alpha \in I\}$ the <u>disjoint union</u> is defined by $\cup\{(X_\alpha \times \{\alpha\}) : \alpha \in I\}$ where we "identify" X_α and $X_\alpha \times \{\alpha\}$.

CATEGORIES. A <u>category</u> is a class (not usually a set), whose members are called <u>objects</u>, and such that for each pair of objects (A,B), there is a set denoted by $\text{Hom}(A,B)$ whose members are called <u>morphisms</u> and are denoted $f: A \to B$, such that the following hold:

(i) given morphisms $f: A \to B$ and $g: B \to C$ there is associated uniquely a morphism $g \circ f: A \to C$

(thus, $\circ$ is a mapping from $\mathrm{Hom}(A,B) \times \mathrm{Hom}(B,C)$ to $\mathrm{Hom}(A,C)$ for objects A, B, C).

(ii) for morphisms $f\colon A \to B$, $g\colon B \to C$, $h\colon C \to D$, we have $h \circ (g \circ f) = (h \circ g) \circ f$,

(iii) for each object A there is a unique morphism $1_A\colon A \to A$ such that for $f\colon A \to B$ and $g\colon B \to A$ we have $f \circ 1_A = f$ and $1_A \circ g = g$.

For example, the class of all sets with mappings as morphisms forms a category. The class of groups with homomorphisms is another example.

A morphism $f\colon A \to B$ is called an <u>isomorphism</u> iff there exists $g\colon B \to A$ such that $f \circ g = 1_B$ and $g \circ f = 1_A$. For example in the category of sets and mappings, the isomorphisms are the bijections.

Let $\mathcal{C}$ be a category. An object A of $\mathcal{C}$ is called <u>universally attracting</u> iff there exists a unique morphism of each object of $\mathcal{C}$ into A, and is called <u>universally repelling</u> iff for each object of $\mathcal{C}$ there exists a unique morphism of A into this object.

RELATIONS. Let $\{X_\alpha\colon \alpha \in I\}$ be a collection of sets. The <u>product</u> is defined by

$$\Pi\{X_\alpha\} = \{f\colon I \to \bigcup\{X_\alpha\colon \alpha \in I\}\colon f(\alpha) \in X_\alpha\}$$

For two sets the product is identified with the <u>cartesian product</u> $X \times Y = \{(x,y)\colon x \in X$ and $y \in Y\}$. If $X_\alpha = X$ for all $\alpha \in I$, we write $X^I = \Pi\{X_\alpha\}$. For example we have the <u>cube</u> $[0,1]^I$.

For a product $Y = \Pi\{X_\alpha\}$ the maps $\pi_\alpha\colon Y \to X_\alpha$ defined by $\pi_\alpha(f) = f(\alpha)$ are the <u>projections</u>. We often identify $y \in Y$ with its projections, or <u>components</u> $y_\alpha = \pi_\alpha(y)$.

Let E be a set. A <u>relation</u> on E is a subset $R \subset E \times E$. We write xRy or $x \sim y$ iff $(x,y) \in R$. A relation R is an <u>equivalence relation</u> iff (i) xRx for all $x \in E$, (ii) $(xRy) \Rightarrow (yRx)$ and (iii) $(xRy$ and $yRz) \Rightarrow (xRz)$.

For a relation R and $x \in E$ we let $R(x) = \{y \in E\colon xRy\}$ and in case R is an equivalence relation we let $R(x) = [x]$, called the <u>equivalence class</u> of x, in which case E is the disjoint union of equivalence classes, and we let E/R denote this set of equivalence classes. The identity relation on E is $\Delta = \{(x,x): x \in E\}$, the <u>diagonal</u> of E.

For $R \subset E \times F$ we say that R is a <u>relation from</u> E <u>to</u> F. (Functions $f: E \to F$ are identified with relations by means of their <u>graph</u>, $\Gamma(f) = \{(x,f(x)); x \in E\}$.) For relations $U \subset E \times F$, $V \subset F \times G$, we define

$$V \circ U = \{(x,z) \in E \times G: \text{for some } y \in F, \\ (x,y) \in U \text{ and } (y,z) \in V\}$$

called the <u>composite relation</u>, and

$$U^{-1} = \{(y,x): (x,y) \in U\}$$

called the <u>inverse relation</u>.

ORDERINGS. Let E be a set. A relation $\leq$ on E is called a <u>partial ordering</u> iff

(i) $(x = y) \iff (x \leq y$ and $y \leq x)$ and

(ii) $(x \leq y$ and $y \leq z) \Rightarrow (x \leq z)$. A relation $\leq$ is a <u>total ordering</u> iff it is a partial ordering and for each $x,y \in E$ either $x \leq y$ or $y \leq x$. A relation is a <u>directed ordering</u> iff it is a partial ordering and for $x,y \in E$ there exists $z \in E$

such that $x \leq z$ and $y \leq z$.

As usual we write $x \geq y$ iff $y \leq x$ and $x < y$ iff $x \leq y$ and $x \neq y$ etc. An example of a directed set is the collection of subsets of a set S with $A \leq B$ iff $A \subset B$. This ordering is not total.

Let E be a partially ordered set. An element $x \in E$ is called <u>maximal</u> iff $x \leq y$ implies $x = y$. We say that $x \in E$ is a <u>greatest element</u> iff $y \leq x$ for all $y \in E$. Greatest elements are maximal but not conversely. Minimal and least elements are defined similarly. For $A \subset E$, we say that $x \in E$ is an <u>upper bound</u> of A iff $a \in A$ implies $a \leq x$, and is the <u>supremum</u>, or <u>least upper bound</u> of A iff x is an upper bound and (y an upper bound) $\Rightarrow$ ($x \leq y$). We denote the supremum by $\sup A$. Clearly $\sup A$ is unique if it exists. Greatest lower bounds are defined similarly. A <u>lattice</u> is a partially ordered set E for which $\inf\{x,y\}$ and $\sup\{x,y\}$ exist for each pair $x,y \in E$. The lattice is <u>complete</u> iff $\sup A$ and $\inf A$ exist for any $A \subset E$, $A \neq \emptyset$. We write $x \vee y = \sup\{x,y\}$ and $x \wedge y = \inf\{x,y\}$.

PROJECTIVE AND INDUCTIVE LIMITS. Let $\mathcal{C}$ be a category and $\{A_\alpha : \alpha \in D\}$ a family of objects of $\mathcal{C}$ for a directed set D. The family, together with morphisms $f_{ij}: A_i \to A_j$, $i \geq j$ is called a <u>projective system</u> iff $f_{jk} \circ f_{ij} = f_{ik}$ and $f_{ii} = 1_A$, for all $i \geq j \geq k$. An object A together with morphisms $f_i: A \to A_i$ such that $f_j = f_{ij} \circ f_i$ is called the <u>projective limit</u> (unique if it exists) of the system iff $(A, \{f_i\})$ is universally attracting in the category of such pairs; that is, for any other pair $(B, \{g_i\})$ there is a unique morphism $h: B \to A$ such that the diagram in figure 1.1 commutes.

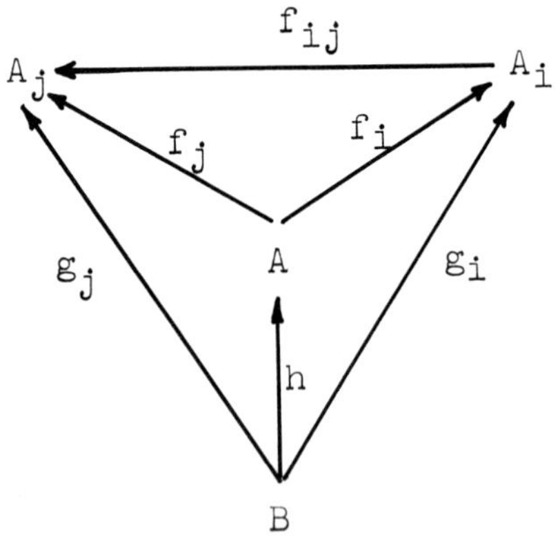

<u>Figure</u> 1.1

For example, if $\{X_\alpha : \alpha \in I\}$ is a family of sets, the product $X = \Pi X_\alpha$ is the projective limit in the category of sets and mappings of the projective system $\Pi\{X_\alpha : \alpha \in F\} = A_F$ for each finite set $F \subset I$ and the F directed by $F \geq G$ iff $F \supset G$, with $\pi_{FG}: A_F \to A_G$ the ordinary projection.

Inductive limits are defined by reversing all the arrows in the definition of projective limits. The reader unfamiliar with inductive limits should write this definition out.

TRANSFINITE INDUCTION AND ZORN'S LEMMA. A partially ordered set E is called _well ordered_ iff every non empty subset has a least element (in the subset). In particular, a well ordered set is totally ordered.

A partially ordered set E is called _inductive_ iff every totally ordered subset has an upper bound in E.

The following assertions are equivalent and will be assumed as part of the axiomatics:

PRINCIPLE OF TRANSFINITE INDUCTION. _Let_ E _be a_

well ordered set and $B \subset E$ satisfy the condition: for each $a \in B$, $(\{x \in E: x < a\} \subset B) \Rightarrow (a \in B)$. Then $B = E$.

AXIOM OF CHOICE. For any family $\{X_\alpha : \alpha \in I\}$ of non-empty sets, $I \neq \emptyset$, there is a mapping $f: I \to (\bigcup\{X_\alpha : \alpha \in I\})$ such that $f(\alpha) \in X_\alpha$.

WELL ORDERING PRINCIPLE. Any set can be well ordered.

ZORN'S LEMMA. If E is partially ordered and inductive then for each $x \in E$ there exists a maximal element $y \in E$ such that $x \leq y$.

CARDINALITY. Two sets E and F are called equipotent iff they are isomorphic in the category of sets. This is an equivalence relation (in a generalized sense) in this category, and the equivalence class of E is denoted card E. We write card $E \leq$ card F iff E is isomorphic to a subset of F. This ordering is partial (Schroder-Bernstein theorem) and in fact total (by Zorn's lemma). Also, card $E <$ card $\mathcal{P}(E)$; we often write card $\mathcal{P}(E) = 2^{\text{card } E}$.

ORDINAL NUMBERS. Consider the category of well ordered sets and order preserving (monotone) maps. Isomorphism in this category defines equivalence classes, called <u>ordinal numbers</u>. The equivalence class of E is denoted Ord E.

For E well ordered, a subset F of E is called <u>initial</u> iff $(x \in F$ and $y \leq x) \Rightarrow (y \in F)$. We say Ord $E \leq$ Ord F iff E is isomorphic to an initial subset of F. Again this order is a total ordering. In fact the ordinal numbers ordered this way are well ordered and we have

$$\text{Ord } E = \text{Ord } \{\text{Ord } F\text{: Ord } F < \text{Ord } E\}.$$

For an ordinal γ, we define its <u>successor</u> by

$$\gamma + 1 = \text{Ord } \{\lambda\text{: } \lambda \leq \gamma\} .$$

Generally in a well ordered set E, $x + 1 = \inf\{y \in E\text{: } x < y\}$. Using initial segments, any well ordered set may be considered as a set of ordinal numbers.

VECTOR SPACES. Let E be a real vector space. A family $\{x_i : i \in I\}$ of elements of E is a <u>Hamel basis</u> of E iff every finite subset is linearly independent and the vector space spanned by (generated by) $\{x_i\}$ is E. Every set of linearly independent elements is a subset of a Hamel basis and any two bases have the same cardinality.

Let $f: E \to \mathbf{R}$ be a linear map. The kernel of f, $f^{-1}(0)$ is called a <u>hyperplane</u> of E. Alternatively, a subset $H \subset E$ is a <u>hyperplane</u> iff it is a subspace and there is a one dimensional subspace $A \subset E$ such that $E = H + A$.

Of course vector spaces and linear maps form a category.

Let $\{E_i : i \in I\}$ be a family of vector spaces. The product $E = \Pi\{E_i\}$ is a vector space if we define $(f+g)(i) = f(i) + g(i) \in E_i$ and $(\alpha f)(i) = \alpha(f(i))$. The subspace consisting of those f for which $f(i) = 0$ for all but finitely many indices i is called the <u>direct sum</u> and is denoted $\Sigma\{E_i\}$. In particular if $E_i = \mathbf{R}$ for all $i \in I$, the product is denoted $\mathbf{R}^I$ and the direct sum by $\mathbf{R}^{(I)}$.

REFERENCES: Lang [2] Algebra

Kamke [1] Theory of Sets

Dugundji [1] Topology

§2 TOPOLOGY AND ANALYSIS

OPEN AND CLOSED SETS. A _topological space_ consists of a set X and a set $\mathfrak{T} \subset \mathcal{P}(X)$ such that $\emptyset$, $X \in \mathfrak{T}$, $\mathfrak{T}$ is closed under finite intersections and arbitrary unions. Elements of $\mathfrak{T}$ are called _open sets_. A _neighborhood_ of $x \in X$ is a set A such that $x \in A$ and there is an open set $U \in \mathfrak{T}$ such that $x \in U \subset A$. A _neighborhood_ of a set $B \subset X$ is a set which is a neighborhood of each point of B; that is, contains an open set containing B. A point $x \in X$ is an _accumulation point_ of $B \subset X$ iff every neighborhood of x contains points of B other than x. A set is _closed_ iff it contains all its accumulation points. Then $B \subset X$ is closed iff $X \setminus B$ is open. A set B together with its accumulation points is called its _closure_ and is denoted $c\ell(B)$; $c\ell(B) = \bigcap\{A \supset B : A \text{ is closed}\}$ and is a closed set since the intersection of any family of closed sets is closed. Thus B is closed iff $B = c\ell(B)$.

The _interior_ of a set A is the set of all points (_interior points_) which have a neighborhood contained in A ; and is denoted $\mathrm{int}(A)$. Thus $\mathrm{int}(A) = \bigcup\{U \subset A\colon U \text{ is open}\}$, and A is open iff $A = \mathrm{int}(A)$. If $A \subset X$ and $c\ell(A) = X$ then A is called <u>dense</u> or <u>everywhere dense</u> in X . If $\mathrm{int}(c\ell(A)) = \emptyset$ then A is called <u>nowhere dense</u>. We call X <u>separable</u> iff X contains a countable subset which is dense in X .

The <u>boundary</u> of a set A denoted $\mathrm{bd}(A)$ consists of accumulation points of both A and $X \setminus A$. Thus $\mathrm{bd}(A) = c\ell(A) \cap c\ell(X \setminus A)$. If A is open, $\mathrm{bd}(A)$ consists of accumulation points of A which do not lie in A .

SEPARATION AXIOMS. A topological space is called T_0 iff for each $x, y \in X$ there is a neighborhood of one of x or y which does not contain the other point. A space is T_1 iff for each $x,y \in X$ there are neighborhoods of each point which do not contain the other point. Equivalently, a space is T_1 iff for each $x \in X$, $\{x\}$ is closed. A space is <u>Hausdorff</u> (or T_2) iff for $x,y \in X$, x and y have disjoint neighborhoods. A space is <u>regular</u> iff for each $x \in X$ and neighborhood V of x there is a closed neighborhood U of x such that

$U \subset V$. A space is _normal_ iff any two disjoint closed sets have disjoint neighborhoods. A space is T_3[resp. T_4] iff it is regular [resp. normal] and T_1 . Obviously, $T_4 \Rightarrow T_3 \Rightarrow T_2 \Rightarrow T_1 \Rightarrow T_0$.

TOPOLOGICAL CATEGORY. Let X and Y be topological spaces and $f: X \to Y$ a mapping. We call f _continuous at_ $x \in X$ iff for every neighborhood U of $f(x)$, $f^{-1}(U)$ is a neighborhood of x . If f is continuous at every $x \in X$ it is called _continuous_. Then f is continuous iff for every open [resp. closed] set $A \subset Y$, $f^{-1}(A)$ is open [resp. closed].

Topological spaces together with continuous mappings form a category (the composition of continuous functions is continuous) called the _topological category_. Isomorphisms in this category are called _homeomorphisms_.

COMPARISON OF TOPOLOGIES. Let $\mathfrak{T}_1$ and $\mathfrak{T}_2$ be two topologies on a set X (so that $(X,\mathfrak{T}_1)$ and $(X,\mathfrak{T}_2)$ are different topological spaces). We say that $\mathfrak{T}_1$ is _finer_ than $\mathfrak{T}_2$ iff $\mathfrak{T}_1 \supset \mathfrak{T}_2$, (or that $\mathfrak{T}_2$ is _coarser_ than $\mathfrak{T}_1$). This is equivalent to continuity of the identity map $i: (X,\mathfrak{T}_1) \to (X,\mathfrak{T}_2)$. Thus the topologies on X form a partially ordered set. The largest element (upper bound) of this set is the <u>discrete topology</u> ($\mathfrak{T} = \mathcal{P}(X)$) and the least element is the <u>trivial</u> (<u>indiscrete</u>) <u>topology</u> ($\mathfrak{T} = \{\emptyset, X\}$).

If $\{\mathfrak{T}_i: i \in I\}$ is a family of topologies on X, then $\mathfrak{T} = \bigcap\{\mathfrak{T}_i\}$ is the greatest lower bound of the $\{\mathfrak{T}_i\}$. Similarly there exists a least upper bound $\mathfrak{T}$ given by: $U \in \mathfrak{T}$ iff U is a union of finite intersections of elements of $\bigcup\{\mathfrak{T}_i\}$, ($\mathfrak{T} \neq \bigcup\{\mathfrak{T}_i\}$).

SUBBASE AND BASE OF A TOPOLOGY. Let X be a set and $\mathfrak{B} \subset \mathcal{P}(X)$. Then the intersection of all topologies containing $\mathfrak{B}$ is called the topology <u>generated</u> by $\mathfrak{B}$. We call $\mathfrak{B}$ a <u>subbase</u> for that topology. A <u>base</u> $\mathfrak{B} \subset \mathcal{P}(X)$ has the property that A, $B \in \mathfrak{B}$ implies $A \cap B$ is a union of elements of $\mathfrak{B}$. The topology generated by a base consists of all unions of elements of the base, and we say it is a <u>base</u> for that topology.

Let $(X,\mathfrak{T})$ be a topological space. A family of neighborhoods of a set is a <u>base for the</u>

neighborhood system of the set iff every neighborhood of the set contains a member of the family. Thus X is regular iff the family of closed neighborhoods of each point is a base for the neighborhood system of the point.

A topological space is called first countable iff there is a countable base for the neighborhood system of each point, and is called second countable iff there is a countable base (or subbase) for the topology. Second countable spaces are separable (but in general the converse doesn't hold).

RELATIVE AND PRODUCT TOPOLOGIES. Let $(X,\mathcal{T})$ be a topological space, and $A \subset X$. Let $\mathcal{A} = \{U \cap A : U \in \mathcal{T}\}$ which we regard as a subset of $\mathcal{P}(A)$. Then $\mathcal{A}$ is a topology on A called the relative topology. When using subsets this topology is understood unless otherwise specified.

Let $(X_i, \mathcal{T}_i)$ be a family of topological spaces and $X = \Pi\{X_i\}$ the product. The product topology is the coarsest topology for which each projection map $\pi_i : X \to X_i$ is continuous. A base for the topology is given by sets of the form

$$\pi_{i_1}^{-1}(U_1) \cap \dots \cap \pi_{i_n}^{-1}(U_n) ,$$

where $U_k \in \mathcal{T}_{i_k}$. (Another but less useful topology for X is the box topology in which we take arbitrary intersections rather than finite ones).

METRIC SPACES. A pseudo-metric space is a set X and a map $d: X \times X \to \mathbb{R}$ such that $d(x,y) \geq 0$, $d(x,y) = d(y,x)$, $d(x,x) = 0$ and $d(x,y) \leq d(x,z) + d(z,y)$ (triangle inequality). If, in addition $d(x,y) = 0$ implies $x = y$ then d is called a metric. A generalized pseudo-metric (or metric) satisfies these conditions except that we allow $d(x,y)$ to be $+\infty$.

Let d be a (generalized pseudo)-metric on X . The (generalized pseudo)-metric topology on X is generated by the base

$$D(x,\varepsilon) = \{y \in X: d(x,y) < \varepsilon\}$$

for $x \in X$ and $\varepsilon > 0$. These ε-discs $D(x,\varepsilon)$ are open. (Generalized pseudo)-metric spaces are first countable. A pseudo-metric space is second countable iff it is separable, and a pseudo-metric space is a metric space iff the topology is Hausdorff.

Let (X,d) and (Y,ρ) be pseudo-metric spaces. A map $f: X \to Y$ is continuous at $x \in X$ iff for every $\varepsilon > 0$ there exists a $\delta > 0$ such that $d(y,x) < \delta$ implies $\rho(f(x), f(y)) < \varepsilon$. The oscillation of a map $f: X \to Y$ in a neighborhood U of x_0 is defined by $\text{oss}(f,U) = \sup\{\rho(f(x), f(y)): x \in U, y \in U\}$, and the oscillation of f at x is given by

$$\text{oss}(f,x_0) = \lim_{\varepsilon \to 0} \text{oss}(f, D(x,\varepsilon))$$

Then f is continuous at x_0 iff the oscillation of f at x_0 is zero. A map $f: X \to Y$ is uniformly continuous iff for any $\varepsilon > 0$ there is a $\delta > 0$ such that $d(x,y) < \delta$ implies $\rho(f(x), f(y)) < \varepsilon$. Also, f is called Lipschitz iff there is a constant $L \geq 0$ such that $\rho(f(x), f(y)) \leq L\, d(x,y)$, for all $x,y \in X$. Obviously Lipschitz maps are uniformly continuous.

The category of metric spaces consists of metric spaces and metric preserving maps. Isomorphisms in this category are called isometries.

COMPLETION. Let $(X,\mathfrak{T})$ be a topological space and x_n a sequence in X. We say that x_n _converges to_ x for $x \in X$ and write $x_n \to x$ iff for every neighborhood U of x there exists an N such that $n \geq N$ implies $x_n \in U$.

Let (X,d) be a (pseudo)-metric space. A _Cauchy sequence_ x_n is a sequence such that for every $\varepsilon > 0$ there is an N such that $n, m \geq N$ implies $d(x_n, x_m) < \varepsilon$. Convergent sequences are Cauchy. We call X _complete_ iff every Cauchy sequence converges (to some point in X).

A _completion_ of X is a complete (pseudo)-metric space (Y,ρ) and a metric preserving injection $\varphi: X \to Y$ such that $\varphi(X)$ is dense in Y.

Completions of a (pseudo)-metric space exist and are all isometric.

Let (X,d) be a metric space and $A \subset X$. The restriction of d to A induces a metric on A and the metric topology coincides with the relative topology. If X is a complete metric space then $A \subset X$ is complete iff A is closed.

Let (Y,ρ) be a complete metric space and (X,d) a pseudo-metric space. Let $A \subset X$ and

$F: A \to Y$ be uniformly continuous. Then there is a unique extension $g: c\ell(A) \to Y$ of f which is also uniformly continuous.

A topological space is (pseudo)-metrizable iff there exists a (pseudo)-metric which induces the given topology.

TYCHONOFF'S LEMMA. Let $(X, \mathfrak{T})$ be a topological space. An open cover of X is a set $\mathscr{A} \subset \mathfrak{T}$ such that $X = \{U: U \in \mathscr{A}\}$. A subcover of $\mathscr{A}$ is a subset $\mathscr{R} \subset \mathscr{A}$ which is also a cover. A space is called Lindelöf iff every open cover has a countable subcover. Lindelöf's theorem (whose proof is trivial), states that every second countable space is a Lindelöf space. A slightly harder theorem is Tychonoff's lemma: each regular Lindelöf space is normal. These notions will be important when we discuss metrization theorems in section 6.

COMPACT SPACES. A topological space $(X, \mathfrak{T})$ is called compact iff every open covering of X has a finite subcovering. A subset is compact iff it is compact in the relative topology. We call $A \subset X$ relatively compact iff $c\ell(A)$ is compact.

Clearly X is compact iff every family of closed sets in X with empty intersection has a finite subfamily whose intersection is also empty.

If $(X,\mathcal{T})$ is a Hausdorff space and $A \subset X$ is compact, then A is closed. A closed subset of a compact space is compact, and a continuous image of a compact space is compact. Let X and Y be compact Hausdorff spaces and $f: X \to Y$ continuous and bijective. Then f is a homeomorphism. In particular, any two comparable compact Hausdorff topologies on X are identical. A compact metric space is second countable (separable).

A metric space X is called <u>totally bounded</u> iff for every $\varepsilon > 0$ there are points $x_1,\ldots,x_n \in X$ such that the sets $D(x_i,\varepsilon)$ cover X. This is equivalent to the assertion that every sequence in X has a Cauchy subsequence. A metric space is compact iff it is complete and is totally bounded. In particular, compact metric spaces are complete. A subset of Euclidean space $\mathbb{R}^n = \mathbb{R} \times \ldots \times \mathbb{R}$ is compact iff it is closed and bounded.

A topological space is called <u>locally compact</u>

iff each point has an open neighborhood which is relatively compact. A locally compact Hausdorff space is regular. If X is locally compact Hausdorff with countable base then X is normal (by Tychonoff's lemma) and there is a sequence of compact sets F_n such that $F_n \subset \mathrm{int}(F_{n+1})$ and $X = \cup\{F_n\}$. A space X is called σ-<u>compact</u> iff there is a sequence of compact sets F_n such that $X = \cup\{F_n\}$.

A <u>compactification</u> of a space X is a compact space Y and a homeomorphism of X onto a dense subspace of Y. This simplest kind of compactification, called the <u>one point compactification</u>, is defined for non-compact spaces as follows. Let $\{\infty\}$ be a set with one point and Y be the disjoint union of X and $\{\infty\}$. The open sets of Y are taken to be the open sets of X and the complements of compact sets in X. Then Y is Hausdorff iff X is locally compact and Hausdorff. For example the one point compactification of $\mathbf{R}^n$ is the sphere $S^n \subset \mathbf{R}^{n+1}$.

CONNECTED SPACES. A topological space X is <u>connected</u> if $\emptyset$ and X are the only sets which are both open and closed. Equivalently, X cannot be written as the disjoint union of two non empty open [resp. closed] sets.

A <u>component</u> of X is a non empty subset which is connected and is not properly contained in any connected set.

A <u>continuum</u> is a compact connected space.

NORMED SPACES. A real (or complex) <u>semi-normed</u> space is a vector space E together with a map $\|\cdot\|: E \to \mathbb{R}$ such that $\|\lambda x\| = |\lambda| \, \|x\|$ and $\|x+y\| \leq \|x\| + \|y\|$. If, in addition $\|x\| = 0$ implies $x = 0$, $\|\cdot\|$ is called a <u>norm</u>. If $\|\cdot\|$ is a semi-norm then $d(x,y) = \|x-y\|$ is a pseudo-metric, and is a metric iff $\|\cdot\|$ is a norm. A normed space is called a <u>Banach space</u> iff it is complete in the induced metric. Let E and F be normed spaces and $f: E \to F$ a linear map. Then f is continuous iff there is a constant μ such that $\|f(x)\| \leq \mu \, \|x\|$ for all $x \in E$. The least upper bound of such constants is called the <u>norm</u> of f, denoted $\|f\|$, and is often called the <u>uniform norm</u>. The collection of continuous linear maps from E to F with the uniform norm is denoted $\mathcal{L}(E,F)$ and is a normed space which is a Banach space if F is a Banach space. In particular we let $E' = \mathcal{L}(E,\mathbb{R})$ if E is real, and $\mathcal{L}(E,\mathbb{C})$ if E is complex, called the <u>dual</u> of E. This is not to be confused with E^* consisting of all linear maps from E to $\mathbb{R}$ (or $\mathbb{C}$).

An <u>inner product</u> on a vector space E (real or complex) is a map: $\langle , \rangle : E \times E \to \mathbb{R}$ (or $\mathbb{C}$) such that $\langle x,x\rangle \geq 0$ and $\langle x,x\rangle = 0$ iff $x = 0$, $\langle x,y\rangle = \langle y,x\rangle^*$ (where $\langle y,x\rangle^*$, denotes conjugate), and $\langle \lambda x + \mu y,z\rangle = \lambda\langle x,z\rangle + \mu\langle y,z\rangle$. If we let $\|x\| = (\langle x,x\rangle)^{1/2}$, then E becomes a normed space.

A <u>Hilbert space</u> is a Banach space whose norm arises from an inner product.

SPACES OF CONTINUOUS FUNCTIONS. Let $(X,\mathcal{J})$ be a topological space and $\mathcal{C}(X,\mathbb{R})$ the vector space of real valued continuous functions on X which are bounded (here $\mathbb{R}$ may be replaced by any Banach space). If we let $\|f\| = \sup\{|f(x)| : x \in X\}$ then $\mathcal{C}(X,\mathbb{R})$ becomes a Banach space, and its norm is called the <u>uniform norm</u>. (If X is compact, every continuous function on X is bounded).

Also, we let $\mathcal{C}_0(X,\mathbb{R})$ denote those functions in $\mathcal{C}(X,\mathbf{R})$ which vanish at infinity (for any $\varepsilon > 0$ there is a compact K outside of which $|f(x)| < \varepsilon$), which is also a Banach space. Finally, we let $\mathcal{K}(X,\mathbf{R})$ denote the elements of $\mathcal{C}(X,\mathbf{R})$ which have compact support (vanish outside some compact.)

Each of $\mathcal{C}$, $\mathcal{C}_o$ and $\mathcal{K}$ are algebras; that is, are vector spaces and closed under (pointwise) products. They are also lattices ordered by $f \geq g$ iff $f(x) \leq g(x)$ for all $x \in X$. (Here we have $f \vee g(x) = \sup\{f(x), g(x)\}$.)

THE STONE-WEIERSTRASS THEOREM. We shall need several forms of this theorem:

(i) Compact Form. Let X be a compact Hausdorff space and $\mathbf{B} \subset \mathcal{C}(X,\mathbf{R})$ a subalgebra which contains the constant function 1 and which separates points (for $x_1, x_2 \in X$, $x_1 \neq x_2$ there exists $f \in \mathbf{B}$ such that $f(x_1) \neq f(x_2)$). Then $\mathbf{B}$ is dense in $\mathcal{C}(X,\mathbf{R})$.

(ii) Complex Form. Let X be a compact Hausdorff space and $\mathbf{B} \subset \mathcal{C}(X,\mathbf{C})$ a subalgebra which contains the constant function 1, which separates points and is closed under complex conjugation.

Then $\mathcal{B}$ is dense in $\mathcal{C}(X,\mathbf{R})$.

(iii) Locally Compact Form. Let X be a locally compact Hausdorff space and $\mathcal{B} \subset \mathcal{C}_0(X,\mathbf{R})$ be a subalgebra which separates points and for each $x \in X$ contains a function f with $f(x) \neq 0$. Then $\mathcal{B}$ is dense in $\mathcal{C}_0(X,\mathbf{R})$. (This is a stronger form of (i)).

(iv) Lattice Form. Let X be a compact Hausdorff space and $\mathcal{B} \subset \mathcal{C}(X,\mathbf{R})$ a vector subspace which is a lattice, separates points and which contains the function $f(x) = 1$.

(Alternative to $\mathcal{B}$ containing 1, we can assume for each $x,y \in X$ and $a,b \in \mathbf{R}$ there is an $f \in \mathcal{B}$ such that $f(x) = a$ and $f(y) = b$). Then $\mathcal{B}$ is dense in $\mathcal{C}(X,\mathbf{R})$.

ASCOLI'S THEOREM. Let $(X,\mathcal{T})$ be a compact topological space and (Y,d) a complete metric space. A family $\mathcal{F}$ of continuous maps $f: X \to Y$ is called equicontinuous iff for each $x \in X$, $\varepsilon > 0$ there is a neighborhood U of x such that $y \in U$ implies $d(f(y), f(x)) < \varepsilon$ for all $f \in \mathcal{F}$.

Let $\mathcal{C}(X,Y)$ denote the complete metric space of continuous maps $f: X \to Y$ with

$$d(f,g) = \sup\{d(f(x), g(x)) : x \in X\}$$

Ascoli's theorem is as follows:

Let X be a compact Hausdorff space and (Y,d) a complete metric space. Then a subfamily $\mathfrak{F} \subset \mathcal{C}(X,Y)$ is relatively compact in $\mathcal{C}$ iff $\mathfrak{F}(x) = \{f(x) : f \in \mathfrak{F}\}$ is relatively compact in Y for each $x \in X$ and $\mathfrak{F}$ is equicontinuous.

Many applications of this theorem will be made in later sections. (For more general versions, which we won't need, see Kelley, General Topology.)

DINI'S THEOREM. Let $(X,\mathfrak{T})$ be a topological space. A map $f: X \to \mathbb{R}$ is lower semi-continuous at $x_0 \in X$ iff for any $\lambda \in \mathbb{R}$ such that $f(x_0) > \lambda$, there is a neighborhood U of x_0 such that $f(x) > \lambda$ for all $x \in U$. Upper semi-continuity is defined by reversing the inequalities. We say f is lower semi-continuous if it is lower semi-continuous for every $x \in X$. We make similar definitions for upper semi-continuity. A map $f: X \to \mathbb{R}$ is lower semi-continuous iff for every $\lambda \in \mathbb{R}$, $\{x: f(x) > \lambda\}$ is open iff $\{x: f(x) \leq \lambda\}$

is closed. A map $f\colon X \to \mathbb{R}$ is continuous iff it is both lower and upper semi-continuous.

Let X be a topological space and $\{f_i\colon i \in I\}$ a family of functions from X into $\mathbb{R}$. The family is called <u>filtering increasing</u> iff for each f_i, f_j there is an f_k in the family such that $f_k \geq f_i$ and $f_k \geq f_j$. Dini's theorem asserts that a filtering increasing family of lower semi-continuous functions whose (pointwise) supremum f is a continuous function satisfies the following condition: for every compact set $K \subset X$ and $\varepsilon > 0$ there is an f_k such that $|f_k(x) - f(x)| < \varepsilon$ for all $x \in K$. (The theorem is almost obvious). A similar statement holds for filtering decreasing families. In particular note that if an increasing sequence of lower semi-continuous functions converge pointwise to a continuous function, then the convergence is uniform on every compact.

REFERENCES FOR §2.

Simmons [1] Introduction to Topology and Modern Analysis.

Kelley [1] General Topology

Yosida [1] Functional Analysis

Bourbaki [1] livre III

§3 MEASURE THEORY

There are several approaches to measure theory, each with its own particular advantage. In this section we review the σ-algebra approach which has the advantage of reaching the convergence theorems quickly. An alternative approach (which is equivalent on locally compact spaces), is that of Radon measures (see chapter III). The latter is the most important for our purposes since it allows a natural topology on the space of measures (the vague topology). Of course it is important to have both techniques at one's disposal.

MEASURES AND OUTER MEASURES. Let X be a set. A class $\mathcal{A} \subset \mathcal{P}(X)$ is a <u>ring</u> iff $\mathcal{A}$ is closed under finite unions and set theoretic differences. If it is closed under countable unions it is called a σ-<u>ring</u>. A ring containing X is called an <u>algebra</u>. A <u>monotone class</u> is a family of subsets closed countable unions and intersections. If is a σ-ring then $\mathcal{A}$ is a monotone class. For $\mathcal{E} \subset \mathcal{P}(X)$, the intersection of the family of all σ-rings (resp. σ-algebras, monotone classes) containing $\mathcal{E}$ is called the σ-ring (resp. σ-algebra, monotone class) generated by $\mathcal{E}$. If $\mathcal{E}$ is a ring, the monotone class and σ-ring generated by $\mathcal{E}$ are identical.

A measure on a set X is a map $\mu: \mathcal{A} \subset \mathcal{P}(X) \to \mathbf{R}^+ \cup \{\infty\}$ where $\mathbf{R}^+ = \{x \in \mathbf{R}: x \geq 0\}$ and $\mathcal{A}$ is a σ-ring (or σ-algebra) such that $\mu(0) = 0$, and if $A_1, A_2, \ldots \in \mathcal{A}$ is a sequence of disjoint sets, then $\mu(\cup\{A_n\}) = \Sigma\{\mu(A_n)\}$.

Often μ is defined only on an algebra $\mathcal{E}$. Then μ is extended to $\mathcal{P}(X)$ by its outer measure.

$$\mu^*(A) = \inf\{\Sigma\{\mu(E_n)\}: E_1, E_2, \ldots \in \mathcal{E},\ A \subset \cup\{E_n\}\}.$$

The basic properties of μ^* are that it is increasing, countably subadditive $(\mu^*(\cup\{A_n\}) \leq \Sigma\{\mu^*(A_n)\})$ and extends μ. A set E is called measurable iff

$$\mu^*(A) = \mu^*(A \cap E) + \mu^*(A \cap X \setminus E) \quad \text{for all } A \subset X.$$

The measurable sets form a σ-algebra and μ^* restricted to these is a measure extending μ. More generally for a measure $\mu: \mathcal{A} \to \mathbf{R}^+ \cup \{\infty\}$ we call the elements of the σ-ring $\mathscr{A}$ the <u>measurable sets</u>, and the triple $(X,\mathscr{A},\mu)$ is a <u>measure space</u>. For example, let F be an increasing function $F: \mathbf{R} \to \mathbf{R}$ continuous from the left $(F(x) = \sup\{F(t): t < x\})$ and let $\mu([a,b]) = F(b) - F(a)$. Then μ extended (by the above theorem) is the <u>Lebesgue-Stieltjes measure</u> associated with F. Using $F(x) = x$ gives <u>Lebesgue measure</u>.

MEASURABLE FUNCTIONS. Let $(X,\mathscr{A},\mu)$ be a measure space. A function $f: X \to \mathbf{R} \cup \{\pm\infty\}$ is called <u>$\mathscr{A}$-measurable</u> (or <u>μ-measurable</u>) iff for any open interval U of $\mathbf{R} \cup \{\pm\infty\}$ we have $f^{-1}(U) \in \mathscr{A}$.

A property P of points of X is said to hold <u>μ-almost everywhere</u> iff $N = \{x \in X: P(x)$ is false$\}$ is contained in a set of measure zero.

<u>Egoroff's theorem</u> states that if $A \in \mathscr{A}$, $\mu(A) < \infty$ and if f_n is a sequence of $\mathscr{A}$-measurable functions finite μ-almost everywhere (or a.e. for short) that converge a.e. to a finite $\mathscr{A}$-measurable function f, then for each $\varepsilon > 0$ there exists $E \in \mathscr{A}$, $E \subset A$ such that $\mu(A \setminus E) < \varepsilon$ and on E the convergence is uniform. If $f,g: X \to \mathbf{R}$ and

we put

$$d(f,g) = \inf\{r \in \mathbb{R}^+ : \mu\{x \in X: |f(x)-g(x)| > r\} \leq r\}$$

then the measurable functions form a complete pseudo-metric space and convergence in this metric is called <u>convergence in measure</u>. (Similar statements hold for f,g with values in a complete metric space).

A <u>simple function</u> is a measurable function which assumes a finite number of values. Every measurable function which is non-negative is a monotone increasing limit of non-negative simple functions.

INTEGRALS. Let $\varphi = \Sigma a_i 1_{A_i}$ be a simple function on a measure space $(X,\mathscr{A},\mu)$. Here 1_A denotes the <u>characteristic function</u> of A (equalling 1 on A and 0 on $X \setminus A$). Define $\int \varphi \, d\mu = \mu(\varphi) = \Sigma a_i \, \mu(A_i)$. For $f \geq 0$, f measurable, define

$$\mu(f) = \int f d\mu = \sup\{\mu(\varphi): \varphi \text{ is a simple function and } 0 \leq \varphi \leq f\} .$$

(If f is bounded this also equals

$\inf\{\mu(\varphi): \varphi$ is a simple function and $f \le \varphi\}$.)

We say that $f \ge 0$ is <u>integrable</u> iff $\mu(f) < \infty$. For $f: X \to \mathbb{R} \cup \{\pm\infty\}$ write $f = f^+ - f^-$ for $f^+ \ge 0$, $f^- \ge 0$ and define f to be <u>integrable</u> iff $\mu(f^+) < \infty$ and $\mu(f^-) < \infty$ and let $\mu(f) = \mu(f^+) - \mu(f^-)$. Then $|\mu(f)| \le \mu(|f|)$.

CONVERGENCE THEOREMS. Let $(X, \mathscr{S}, \mu)$ be a measure space. The <u>Fatou lemma</u> states that if f_n is a sequence of non-negative measurable functions then

$$\int (\liminf f_n)\, d\mu \le \liminf \int f_n \, d\mu .$$

Recall that

$$\liminf a_n = \lim_{n\to\infty} (\inf\{a_k: k \ge n\}) , \quad \text{and}$$
$$\limsup a_n = \lim_{n\to\infty} (\sup\{a_k: k \ge n\}) .$$

The <u>monotone convergence theorem</u> states that if f_n is an increasing sequence of non-negative measurable functions converging pointwise to f then (f is measurable and) $\int f \, d\mu = \text{limit} \int f_n \, d\mu$.

Finally the <u>Lebesgue dominated convergence</u> theorem states that if f_n is a sequence of measurable functions, $f_n \to f$ a.e. and there is an integrable function g such that $|f_n| \leq g$, then $\int f \, d\mu = \text{limit} \int f_n \, d\mu$.

FUBINI'S THEOREM Let $(X, \mathscr{A}, \mu)$ and $(Y, \mathscr{B}, \nu)$ be measure spaces and $\mathscr{A} \times \mathscr{B}$ denote the σ-algebra generated by $\{A \times B : A \in \mathscr{A}, B \in \mathscr{B}\}$. There is a unique measure on $\mathscr{A} \times \mathscr{B}$ denoted $\mu \times \nu$ such that $(\mu \times \nu)(A \times B) = \mu(A)\nu(B)$ and it is called the <u>product measure</u>. Fubini's theorem states that if $f: X \times Y \to \mathbf{R} \cup \{\pm\infty\}$ is $\mu \times \nu$-integrable, then

$$\int f \, d(\mu \times \nu) = \int_X \left(\int_Y f \, d\nu\right) d\nu = \int_Y \left(\int_X f \, d\mu\right) d\nu$$

where $\left(\int_Y f \, d\nu\right)(x) = \int f_x \, d\nu$; $f_x(y) = f(x,y)$.

If the spaces are σ-finite one can also deduce integrability of f from the existence of the iterated integrals (Tonelli's theorem). (A measure space is σ-<u>finite</u> iff it is the countable union of measurable sets of finite measure.)

THE L_p SPACES Let $(X, \mathscr{A}, \mu)$ be a measure space and $p \in \mathbf{R}$, $1 \leq p < \infty$.

Define

$$\mathcal{L}_p(X,\mathcal{A},\mu) = \{f: X \to \mathbb{R} \cup \{\pm\infty\}: |f|^P \text{ is integrable}\}$$

Then $\mathcal{L}_p$ is a complete semi-normed space with semi-norm $\|f\|_p = (\int |f|^p \, d\mu)^{1/p}$. The triangle inequality is called <u>Minkowski's</u> <u>inequality</u>. If we identify (that is, form equivalence classes) f with g iff $f = g$ a.e. then $\mathcal{L}_p$ becomes a Banach space denoted L_p . If $f \in L_p$ and $g \in L_p$ where $1 < p < \infty$, $1 < q < \infty$ and $1/p + 1/q = 1$ then we have $fg \in L_1$ and

$$|\int fg \, d\mu| \leq \|f\|_p \, \|g\|_q$$

(<u>Holder's</u> <u>inequality</u>). In fact, the map $\varphi: L_p \to L_q'$ given by $\varphi(f)\cdot g = \int fg \, d\mu$ is an isometric isomorphism (onto), (the above inequality shows that $\varphi(f)$ is continuous). In particular for the Hilbert space L_2, $L_2 \approx L_2'$.

Also we let $\mathcal{L}_\infty = \{f: X \to \mathbb{R} \cup \{\pm\infty\}$: there is an $M \geq 0$ such that $|f| \leq M$ a.e.$\}$, and the greatest lower bound of such M is denoted $\|f\|_\infty$. Again $\mathcal{L}_\infty$ is a complete semi-normed space and we can form the associated Banach space L_∞ . We have

$L_1' = L_\infty$ but only $L_\infty' \supset L_1$.

SIGNED MEASURES. Let X be a set. A <u>signed measure</u> is a map $\mu: \mathcal{A} \to \mathbf{R} \cup \{\pm\infty\}$, where $\mathcal{A}$ is a σ-algebra such that μ assumes at most one of the values $\pm\infty$, $\mu(O) = 0$ and if $A_1, A_2, \ldots \in \mathcal{A}$ are disjoint, $\mu(\cup\{A_n\}) = \Sigma\mu(A_n)$.

The <u>Jordan-Hahn decomposition theorem</u> states that if μ is a signed measure, we can write $\mu = \mu^+ - \mu^-$ where μ^+ and μ^- are positive measures on $\mathcal{A}$ and μ^+ and μ^- are <u>disjoint</u>; that is, there are sets $A, B \in \mathcal{A}$, $X = A \cup B$, $A \cap B = \emptyset$ such that $\mu^+(A) = 0$ and $\mu^-(B) = 0$. The positive measure $\mu = \mu^+ + \mu^-$ is called the <u>total variation</u> of μ , or the <u>absolute value</u> of μ .

INTEGRATION ON LOCALLY COMPACT SPACES. Let X be a locally compact Hausdorff space which is also σ-compact (σ-compact is included mainly to avoid confusions in terminology). The <u>Borel sets</u> are elements of the σ-algebra generated by the open (or closed) sets, and are denoted by $\mathcal{B}$.

A <u>Borel measure</u> is a measure defined on $\mathcal{B}$ such that the measure of every compact is finite. The measure is <u>regular</u> iff for every open set U, $\mu(U) = \sup\{\mu(K): K \subset U; K$ is compact$\}$. (This is equivalent to: for every Borel or measurable set A, $\mu(A) = \inf\{\mu(U): A \subset U,\ U$ open$\}$. In this case, for $\mu(A) < \infty$, we have $\mu(A) = \sup\{\mu(K): K$ compact, $K \subset A\}$. If μ is not regular we can obtain a regularization by defining $\nu(A) = \inf\{\mu(U): A \subset U: U$ open$\}$. If two regular Borel measures coincide on the open [resp. compact] sets, they are equal.

A <u>positive Radon measure</u> is a positive linear map $\varphi: \mathcal{K}(E,\mathbf{R}) \to \mathbf{R}$ where $\mathcal{K}(E,\mathbf{R})$ denotes the set of continuous functions with compact support, and φ positive means $f \geq 0 \Rightarrow (\varphi(f) \geq 0)$.

The <u>Riesz representation theorem</u> establishes a one to one correspondence between regular Borel measures and positive Radon measures; for each positive Radon measure there is a unique regular Borel measure μ such that $\varphi(f) = \int f d\mu$ for each $f \in \mathcal{K}(E,\mathbf{R})$.

In this context, if $1 \leq p < \infty$, then $\mathcal{K}(E,\mathbf{R})$ is dense in $\mathcal{L}_p$.

THE DANIELL INTEGRAL. The Daniell integral is rather like a Radon measure in that it begins directly with a positive linear form. It is important however because it can take us from the locally compact case to a more general setting, for example to Hilbert spaces. We shall use Daniell integrals in connection with "conical measures" later in the book.

Let X be a set and E a lattice vector space of functions $f: X \to \mathbf{R}$ (that is, E is a vector space and $f, g \in E$ implies $\inf(f,g) \in E$, where $\inf$ is taken in the usual ordering: $f \geq g$ iff $f(x) \geq g(x)$ for all $x \in X$.) A <u>Daniell integral</u> is a positive linear form $I: E \to \mathbf{R}$, such that if φ_n is an increasing sequence of functions in E and $\varphi \leq \lim \varphi_n$ for some $\varphi \in E$, then $I(\varphi) \leq \lim I(\varphi_n)$. (This is equivalent to: if $\varphi_n \downarrow 0$ then $I(\varphi_n) \downarrow 0$).

Let E_u denote functions $f: X \to \mathbf{R}$ which are increasing limits of functions in E. Then I is extended to E_u by taking limits. For an arbitrary $f: X \to \mathbf{R}$, define the <u>upper integral</u>

$$UI(f) = \inf\{I(g): g \in E_u, f \leq g\}$$

and the <u>lower integral</u>

$$LI(f) = \sup\{I(g): g \in E_u, g \leq f\} .$$

We say f is integrable iff $UI(f) = LI(f)$, and we let $\mathcal{L}_1$ denote the integrable functions.

In this connection one recovers the Fatou lemma and the other convergence theorems. Also, f is called <u>measurable</u> iff $\inf(g,f) \in \mathcal{L}_1$ for all $g \in \mathcal{L}_1$.

The following theorem is fundamental and the hypotheses are quite essential. Namely if I is a Daniell integral on E, a vector lattice of real valued functions on a set X, and if $(f \in E) \Rightarrow (1 \wedge f \in E)$, then there is a measure μ on a σ-algebra $\mathcal{A} \subset \mathcal{P}(X)$ such that $\mathcal{L}_1$ equals the μ-integrable functions and for $f \in \mathcal{L}_1$,

$$I(f) = \int f d\mu .$$

If $\mathcal{A}$ is the smallest σ-algebra such that each I-measurable function is $\mathcal{A}$-measurable and if $1 \in E$, then μ is unique. (This theorem is due to Stone).

REFERENCES. Halmos [2] <u>Measure Theory</u>

Royden [1] <u>Real Analysis</u>

Bourbaki [2] (Livre VI).

CHAPTER 2

SOME TOPOLOGICAL TOOLS IN ANALYSIS

Every analyst should be equipped with a substantial amount of topology beyond those elementary notions reviewed in §2. This chapter is designed to provide some of those tools.

There are two central notions developed in this chapter, namely uniform spaces and Baire spaces. Uniform spaces generalize metric spaces by providing a concept of "uniform" closeness such as is needed to define uniform continuity. Later we shall study topological vector spaces which are not metrizable, but nevertheless have a natural uniform structure. Baire spaces provide a useful way to express the size of a set, analogous to a measure, even though no measure is given. For example, knowing that a topological vector space is a Baire space can give easy proofs of many important theorems for that space.

Standard references for this chapter are Bourbaki (Livre III); Kelley [1] General Topology, and Kuratowski [1] Topology.

§4 NETS AND FILTERS.

In this section we explain the terminology of nets and filters, and discuss how these concepts are used to prove topological theorems. We also discuss initial and final topologies.

We begin with the definition and some examples of filters.

4.1 DEFINITION. Let E be a non empty set and $\mathcal{P}(E)$ the set of subsets of E. A filter on E is a subset $\mathfrak{F} \subset \mathcal{P}(E)$ satisfying the following properties:

(i) $\emptyset \notin \mathfrak{F}$

(ii) $(X, Y \in \mathfrak{F}) \Rightarrow (X \cap Y \in \mathfrak{F})$

(iii) $(X \in \mathfrak{F}$ and $X \subset Y \subset E) \Rightarrow (Y \in \mathfrak{F})$

4.2 EXAMPLES. (i) Suppose E is a topological space and $a \in E$. Let $\mathcal{U}_a$ be the set of all neighborhoods of a. Then $\mathcal{U}_a$ is a filter, called the filter of neighborhoods of a. Note that the open [resp. closed] neighborhoods of a will not in general be a filter, because of 4.1 (iii).

Every filter $\mathfrak{F}$ on a set E with $\bigcap\{X \in \mathfrak{F}\} \neq \emptyset$ is the filter of neighborhoods of $a \in \bigcap\{X \in \mathfrak{F}\}$ if, on E, we put the topology $\mathfrak{T} = \mathfrak{F} \cup \{\emptyset\}$. When $\bigcap\{X \in \mathfrak{F}\} = \emptyset$, we adjoin a point a to E; then $\mathfrak{F}' = \{X \cup \{a\}: X \in \mathfrak{F}\}$ is a filter on $E \cup \{a\}$ with $\bigcap\{Y \in \mathfrak{F}'\} = \{a\}$ and so the previous remark applies to $\mathfrak{F}'$.

(ii) Let E be a set and $a \in E$. The <u>trivial filter</u> of a is defined by $\mathfrak{F}_a = \{X \subset E: a \in X\}$. Then $\mathfrak{F}_a$ is a filter of neighborhoods in some topology on E iff $\{a\}$ is an open set. Thus each $\mathfrak{F}_a$ for $a \in E$ is a filter of neighborhoods in some topology on E iff that topology is discrete.

(iii) Let $\mathbb{N}$ denote the set of non-negative integers; $\mathbb{N} = \{0, 1, 2, \ldots\}$. Let

$$\mathfrak{F} = \{A \subset \mathbb{N}: \mathbb{N} \setminus A \text{ is finite}\}$$

Then $\mathfrak{F}$ is a filter with the property $\bigcap\{X \in \mathfrak{F}\} = \emptyset$. This filter is sometimes called the elementary, or <u>Fréchet filter</u>.

It often occurs that one is not given a filter, but rather a filter base, which is analogous to being given a base of neighborhoods.

We make this precise in the following definition:

4.3 DEFINITION. Let E be a set and $\mathfrak{F}$ a filter on E. A base $\mathfrak{B}$ of $\mathfrak{F}$ is a subset of $\mathfrak{F}$ such that for each $X \in \mathfrak{F}$ there is a $Y \in \mathfrak{B}$ such that $Y \subset X$. (In other words, $\mathfrak{F} = \{X \subset E$: there exists $Y \in \mathfrak{B}$ such that $Y \subset X\}$.)

A filter base on E is a set $\mathfrak{B} \subset \mathcal{P}(E)$ such that (i) $\emptyset \notin \mathfrak{B}$

and (ii) $(X,Y \in \mathfrak{B}) \Rightarrow$ (there exists $Z \in \mathfrak{B}$ such that $Z \subset X \cap Y$).

Thus, if $\mathfrak{B}$ is a filter base then $\mathfrak{B}$ is a base of the filter $\mathfrak{F} = \{X \subset E$: there exists $Y \in \mathfrak{B}$ such that $Y \subset X\}$, called the filter generated by $\mathfrak{B}$.

4.4 EXAMPLES. (i) Let E and F be sets and $f: E \to F$ a mapping. If $\mathfrak{F}$ is a filter on E then $f(\mathfrak{F}) = \{f(X): X \in \mathfrak{F}\}$ is a filter base on F but not necessarily a filter. If f is onto, however, $f(\mathfrak{F})$ is a filter.

(ii) Again let $f\colon E \to F$ be a mapping. If $\mathfrak{F}$ is a filter on F then $f^{-1}(\mathfrak{F}) = \{f^{-1}(X)\colon X \in \mathfrak{F}\}$ need not be a filter base. However, if $f^{-1}(X) \neq \emptyset$ for all $X \in \mathfrak{F}$, then $f^{-1}(\mathfrak{F})$ is a filter base, but not necessarily a filter.

Before proceeding with the discussion of filters, we introduce a closely related concept, that of nets (generalized sequences).

4.5 DEFINITION. Recall that a directed set D is a partially ordered set such that for each m, $n \in D$ there is a $p \in D$ so that $p \geq m$ and $p \geq n$.

Let E be a set. A net on E is a mapping $f\colon D \to E$ where D is a directed set. By analogy with sequences, we generally write x_α, $\alpha \in D$ for a net.

The basic correspondence between nets and filters is the following. Let x_α be a net on a set E. Then $\mathfrak{B} = \{\{x_\alpha\colon \alpha \geq \alpha_0\}\colon \alpha_0 \in D\}$ is a filter base on E . The filter it generates is called the <u>filter of sections</u>. Conversely if $\mathfrak{F}$ is a filter then $\mathfrak{F}$ is a directed set with $(X \leq Y) \iff (Y \subset X)$. An associated net is obtained by selecting $x_X \in X$.

Later, when E has a topology, we shall see how convergence of filters and nets are linked.

For theorems in topology it is sometimes convenient to use filters when nets are clumsy, and vice-versa. Roughly speaking, filters are often more natural when dealing with compact spaces, while nets are convenient for proving a set is closed.

The basic connection between filters and a topological structure is the notion of convergence. On some interesting spaces we have a notion of convergence, but no topology, and again filters are useful. (For example, such spaces arise in differential calculus; see Frolicher and Bucher, [1] <u>Calculus in Vector Spaces Without Norm</u>). For our purposes, however, we consider only the topological case.

4.6 <u>DEFINITION</u>. Let E be a topological space and $\mathcal{F}$ a filter on E. We say that $\mathcal{F}$ <u>converges</u>

to $a \in E$ iff for every neighborhood U of a, $U \in \mathfrak{F}$.

Similarly, a net x_α <u>converges</u> to $a \in E$ iff for every neighborhood U of a, there is an α_o such that $\alpha \geq \alpha_o$ implies $x_\alpha \in U$. In this case we write $x_\alpha \to a$.

A net, or filter, is <u>convergent</u> iff there exists an $a \in E$ to which it converges.

From the definitions we have the following: If $\mathfrak{F}$ is a filter on E, then ($\mathfrak{F}$ converges to $a \in E$) $\Rightarrow$ (the corresponding net x_X, $X \in \mathfrak{F}$, converges to a). Similarly, if x_α is a net in E and $a \in E$ then $x_\alpha \to a$ iff the filter of sections corresponding to x_α together with a, converges to a.

4.7 <u>PROPOSITION</u>. <u>Let</u> E <u>be a topological space</u>. <u>Then</u> $A \subset E$ <u>is closed iff for any net</u> x_α <u>with</u> $x_\alpha \in A$ <u>which converges to</u> $a \in E$, <u>we have</u> $a \in A$.

<u>Proof</u>. If A is closed then $a \in A$, as a is a cluster point of the set $\{x_\alpha : \alpha \in D\}$. Conversely, if A were not closed there would be a

cluster point $a \in E \setminus A$. For U any neighborhood of a , choose $x_U \in U \cap A$. Then x_U forms a net in A converging to a . □

This proposition typifies operations with nets. Another, whose proof we leave to the reader, is the following:

4.8 PROPOSITION. *Let E and F be topological spaces and* $f: E \to F$ *a mapping. Then* (f *is continuous*) $\Leftrightarrow \{(x_\alpha \to a$ *in* $E) \Rightarrow (f(x_\alpha) \to f(a)$ *in* $F)\}$. *If* E *is first countable, nets may be replaced by sequences.*

We return now to some basic properties of filters.

4.9 DEFINITION. Let E be a set. We put on the set of filters the partial ordering of set inclusion; $\mathfrak{F}_1 \leq \mathfrak{F}_2$ iff $\mathfrak{F}_1 \subset \mathfrak{F}_2$, in which case we say that $\mathfrak{F}_2$ is *finer* than $\mathfrak{F}_1$, or $\mathfrak{F}_1$ is *coaser* than $\mathfrak{F}_2$.

An *ultrafilter* on E is a filter $\mathfrak{F}$ on E which is maximal; that is, $\mathfrak{F}_1 \supset \mathfrak{F}$ implies $\mathfrak{F}_1 = \mathfrak{F}$.

For example, in the plane $E = \mathbf{R}^2$, let $\mathfrak{F}_1$ be the filter of neighborhoods of 0 and $\mathfrak{F}_2$ the filter generated by the filter base of neighborhoods of 0 on the x-axis. Then $\mathfrak{F}_2$ is finer than $\mathfrak{F}_1$, but neither is an ultrafilter.

For any set E, and $a \in E$, the trivial filter $\mathfrak{F}_a$ (see 4.2 (ii)) is an ultrafilter. In fact, these are the only "explicitly known" ultrafilters, although others exist; see problem 4.2.

The collection of filters on a set E has a least element, namely $\mathfrak{F} = \{E\}$. However, if E has more than one element, there is no largest filter. [Proof: For $a,b \in E$, $a \neq b$, there is no filter containing both of the trivial filters $\mathfrak{F}_a$ and $\mathfrak{F}_b$.]

Notice that in a topological space E, $\mathfrak{F}$ converges to $a \in E$ iff $\mathfrak{F}$ is finer than the filter of neighborhoods of a.

4.10 PROPOSITION. Let E be a set. Then

(i) every filter on E is contained in some ultrafilter on E

(ii) if $\mathfrak{F}$ is a filter on E, the following are equivalent:

(a) $\mathfrak{F}$ <u>is an ultrafilter</u>

(b) <u>for each</u> $X \in \mathfrak{F}$, <u>if</u> $X = A \cup B$, <u>and</u> $A \cap B = \emptyset$ <u>then</u> $A \in \mathfrak{F}$ <u>or</u> $B \in \mathfrak{F}$

(c) $(E = A \cup B\ ,\ A \cap B = \emptyset) \Rightarrow (A \in \mathfrak{F} \text{ or } B \in \mathfrak{F})$

(d) $(X_1, \ldots, X_n \subset E$ <u>and</u> $E = \bigcup\{X_i\}) \Rightarrow$ $(X_i \in \mathfrak{F}$ <u>for some</u> $i)$.

<u>Proof</u>. (i) The set of ultrafilters on E is inductive so that Zorn's lemma shows that every element is majorized by a maximal one.

(ii) (a) $\Rightarrow$ (b) . We claim that either A or B meets every element of $\mathfrak{F}$. For suppose that $A \cap X_1 = B \cap X_2 = \emptyset$ for some $X_1, X_2 \in \mathfrak{F}$. This would imply $X \cap X_1 \cap X_2 = \emptyset$, a contradiction. Now if A meets every element of $\mathfrak{F}$, then

$$\mathfrak{F} \cup \{A \cap Y\colon Y \in \mathfrak{F}\}$$

is a filter base, which equals $\mathfrak{F}$ as $\mathfrak{F}$ is maximal. Hence $A = A \cap E$ belongs to $\mathfrak{F}$.

(b) $\Rightarrow$ (c). This is obvious as $E \in \mathfrak{F}$.

(c) $\Rightarrow$ (a). Suppose $\mathfrak{F}$ is not maximal. Thus there is a filter $\mathfrak{F}' \supset \mathfrak{F}$ but $\mathfrak{F}' \neq \mathfrak{F}$. If $X \in \mathfrak{F}' \setminus \mathfrak{F}$ then $E \setminus X \in \mathfrak{F}$ by (c). But this means $E \setminus X \in \mathfrak{F}'$, so that $X \cap (E \setminus X) = \emptyset \in \mathfrak{F}'$, a contradiction. We leave the final step (b) <=> (d) to the reader. □

Notice that this proposition holds in particular for the trivial ultrafilters $\mathfrak{F}_a$. Regarding the ordering of filters, we have:

4.11 PROPOSITION. *Let* E *be a set and* $\{\mathfrak{F}_i\}$ *a family of filters on* E . *Then*

(i) *the family has a greatest lower bound, namely the filter* $\cap \{\mathfrak{F}_i\}$

(ii) *the family has a least upper bound iff* $\{\mathfrak{F}_i\}$ *has the finite intersection property; that is, finite intersections of elements of* $\cup \{\mathfrak{F}_i\}$ *are non-empty*.

The proof is immediate. Note that in (ii) the least upper bound is generated by the filter base which consists of finite intersections of elements of $\cup \{\mathfrak{F}_i\}$.

One useful consequence of the characterization of ultrafilters in 4.10 (ii) is:

4.12 PROPOSITION. *Let* E *and* F *be sets and*

$f: E \to F$ _a surjection_. _If_ $\mathfrak{F}$ _is an ultrafilter on_ E _then_ $f(\mathfrak{F})$ _is an ultrafilter on_ F.

Again the proof is immediate.

We turn now to some more connections between filters and a topological structure. Some of these statements may be translated into a corresponding proposition for nets. We leave this task to the reader.

4.13 _PROPOSITION_. _Let_ $\{E_i\}$ _be a family of topological spaces and_ E _be the product space (with the product topology), with_ $\pi_i: E \to E_i$ _the_ i^{th} _projection map_. _Let_ $\mathfrak{F}$ _be a filter on_ E _and_ $\mathfrak{F}_i = \pi_i(\mathfrak{F})$. _Then_ $\mathfrak{F}$ _converges to_ $a \in E$ _iff_ $\mathfrak{F}_i$ _converges to_ $a_i = \pi_i(a)$ _for each_ i.

Proof. If $\mathfrak{F}$ converges to a then $\mathfrak{F}_i$ converges to a_i, for if U_i is a neighborhood of a_i then $\pi_i^{-1}(U_i) \in \mathfrak{F}$, or $U_i \in \mathfrak{F}_i$.

Conversely, if each $\mathfrak{F}_i$ converges to a_i, and U_i is a neighborhood of a_i then $\pi_i^{-1}(U_i) \in \mathfrak{F}$, for if $U_i = \pi_i(V_i) \in \mathfrak{F}_i$ for $V_i \in \mathfrak{F}$, $\pi_i^{-1}(U_i) \supset V_i$. Hence as $\mathfrak{F}$ is closed under finite intersections every basic neighborhood of a belongs to $\mathfrak{F}'$. □

4.14 <u>DEFINITION</u>. Let $\mathfrak{F}$ be a filter on a topological space E. The set of <u>cluster points</u> of $\mathfrak{F}$ is defined by

$$c\ell(\mathfrak{F}) = \bigcap \{c\ell(X) : X \in \mathfrak{F}\} .$$

For example, if $\mathfrak{F}$ converges to $a \in E$ then a is a cluster point of $\mathfrak{F}$. (For if U is a neighborhood of a, $U \cap X \neq \emptyset$ for all $X \in \mathfrak{F}$, as $U \in \mathfrak{F}$; thus $a \in c\ell(X)$.)

We now characterize compact spaces:

4.15 <u>PROPOSITION</u>. <u>Let</u> E <u>be a topological space</u>. <u>Then the following are equivalent</u>:

(i) E <u>is compact</u>

(ii) <u>every filter on</u> E <u>has a cluster point</u>

(iii) <u>for every filter</u> $\mathfrak{F}$, <u>there is a filter</u> $\mathfrak{F}_0$ <u>which converges and</u> $\mathfrak{F}_0 \supset \mathfrak{F}$.

(iv) <u>every ultrafilter on</u> E <u>converges</u>.

<u>Proof</u>. That (i) $\Longleftrightarrow$ (ii) follows from the finite intersection property of compact spaces.

(ii) $\Rightarrow$ (iii). Let $\mathfrak{F}_0$ be an ultrafilter containing $\mathfrak{F}$. Let a be a cluster point of $\mathfrak{F}_0$, and note that $\mathfrak{F}_0$ and $\mathcal{U}_a$, the filter of neighborhoods, have an upper bound by 4.11 (ii). Since $\mathfrak{F}_0$ is maximal, this implies $\mathcal{U}_a \subset \mathfrak{F}_0$, or $\mathfrak{F}_0$ converges to a .

(iii) $\Rightarrow$ (ii) if $\mathfrak{F}_0 \supset \mathfrak{F}$ and $\mathfrak{F}_0$ converges to a , then a is a cluster point of $\mathfrak{F}_0$ and hence of $\mathfrak{F}$.

Finally, it is obvious that (iii) and (iv) are equivalent. □

The next theorem illustrates just how useful filters are.

4.16 THEOREM. (Tychonoff). *If $\{E_i\}$ is a family of compact spaces then the product space E is compact.*

Proof. (Bourbaki). If $\mathfrak{F}$ is an ultrafilter on E then $\mathfrak{F}_i = \pi_i(\mathfrak{F})$ is an ultrafilter on E_i by 4.12. But each $\mathfrak{F}_i$ converges (4.15), so that $\mathfrak{F}$ converges (4.13). Hence E is compact. □

There is an important way of generating

topologies on a set, which is the following:

4.17 *DEFINITION*. Let E be a set, $\{E_i\}$ a family of topological spaces and $f_i: E \to E_i$ mappings. The *initial* (or *projective*) topology on E is the coarsest in which each f_i is continuous.

Dually, for mappings $g_i: E_i \to E$ the *final* (or *inductive*) topology on E is the finest topology in which each g_i is continuous.

These topologies exist. The initial topology is generated by $\{f_i^{-1}(U): U \subset E_i \text{ open}\}$, while the final topology is given by

$$\{U \subset E: f_i^{-1}(U) \text{ is open in } E_i \text{ for all } i\}.$$

The names projective and inductive topologies are well chosen for they are the topologies associated with projective and inductive limits; see problem 4.3.

4.18 *PROPOSITION*. (i) *Let* E *have the initial topology with respect to maps* $f_i: E \to E_i$. *For a topological space* F , *a map* $g: F \to E$ *is continuous iff* $f_i \circ g$ *is continuous for each* i .

(ii) <u>Let</u> E <u>have the final topology with respect to maps</u> $g_i\colon E_i \to E$. <u>Then a map</u> $f\colon E \to F$ <u>is continuous iff</u> $f \circ g_i$ <u>is continuous for all</u> i.

<u>Proof</u>. (i) If g is continuous then obviously $f_i \circ g$ is continuous. Conversely if each $f_i \circ g$ is continuous and $U_i \subset E_i$ is open then $g^{-1}(f_i^{-1}(U_i))$ is open. Hence g is continuous, as $\{f_i^{-1}(U_i)\}$ generates the topology on E .

The proof of (ii) is even easier and is left to the reader. □

An important example of a final topology is the <u>quotient</u> <u>topology</u>. Namely, if E is a topological space and R is an equivalence relation on E , then the final topology on E/R (the set of equivalence classes) with respect to the canonical projection $\pi\colon E \to E/R$ is called the quotient topology.

4.19 <u>PROPOSITION</u>. <u>Let</u> E <u>be a set</u>, $\{E_i\}$ <u>a family of topological spaces and</u> $f_i\colon E \to E_i$ <u>mappings</u>. <u>Let</u> F <u>denote the product of the</u> E_i <u>and define</u>

$f: E \to F$ by $(f(a))_i = f_i(a)$, (*the* i^{th} *coordinate*).

(i) *the initial topology on* E *relative to the* f_i *coincides with the initial topology relative to* f .

(ii) f *is one to one iff* $(a \neq b \in E) \Rightarrow$ (*there exists* i *such that* $f_i(a) \neq f_i(b)$) .

(iii) *if* f *is one to one and* E *has the initial topology then* $f: E \to f(E)$ *is a homeomorphism*, $f(E)$ *having the relative topology*.

(iv) *if* f *is one to one and the* E_i *are compact*, *then* E *is homeomorphic to a dense subset of a compact space, namely* $c\ell(f(E))$. *That is*, E *has a compactification*.

Proof. (i) This is obvious from the definitions. (The topology in each case is generated by the sets $\{f_i^{-1}(U_i)\}$ for $U_i \subset E_i$ open).

(ii) is clear

(iii) follows from (i) and the fact that if $f: E \to H$ is a bijection and E has the initial topology, then f is a homeomorphism (obvious).

Finally (iv) follows from (iii). □

A typical application is the following. Suppose the topology on E is generated by a separating family of maps $f_i: E \to \mathbb{R}$. Then E may be embedded in $\mathbb{R}^I$; $i \in I$. If the maps f_i are bounded, E has a compactification.

PROBLEMS FOR §4.

4.1 Countably Accessible Nets.

A net x_α, $\alpha \in D$ is called countably accessible iff D has a countable cofinal subset (has arbitrarily large elements).

Show that a countably accessible net x_α in a Hausdorff space converges iff every countable subnet, obtained from cofinal subsets of D, converges.

Give an example.

4.2 Non-trivial Ultrafilters. Prove that non-trivial ultrafilters exist on $\mathbb{R}$. Generalize.

4.3 Projective and Inductive Limits. Prove that projective and inductive limits exist in the topological category.

4.4 Extreme Measures and Ultrafilters.

Let A be a (non-empty) set and U be the set of ultrafilters on A.

A mapping $f: \mathcal{P}(A) \to \mathbb{R}$ is called a (finitely additive) measure iff $f(X) \geq 0$ for all $X \subset A$ and $(X \cap Y = \emptyset) \Rightarrow f(X \cup Y) = f(X) + f(Y)$. Also, f is called extremal iff $f(X) = 0$ or $f(X) = 1$ for all $X \subset A$ and $f(A) = 1$.

For a measure f, let

$$u(f) = \{X \subset A: f(X) = 1\}$$

(i) Show that if f is extremal then $u(f)$ is an ultra-filter on A. Show conversely that every ultrafilter on A arises from a unique extremal measure this way.

(ii) Let E denote the set of extremal measures on A with the topology of simple convergence (the initial topology for the maps $f \mapsto f(X) \in \mathbb{R}$ for each $X \subset A$). On U put the topology which makes $u: E \to U$ a homeomorphism. Show that E, or U, is compact.

(iii) For each $X \subset A$, $X \neq \emptyset$, let $\omega(X)$ denote the set of ultrafilters on A which contain

X , and let $\omega(\emptyset) = \emptyset$. Show that $\omega(X)$ is both open and closed in U and that $\{\omega(X): X \subset A\}$ is a base of open sets of U . Characterize all the subsets of U which are both open and closed and determine the isolated points of U .

(iv) A <u>positive Radon measure</u> on U is a linear map $\mu: \mathcal{C}(U,\mathbf{R}) \to \mathbf{R}$ such that $(\varphi \geq 0) \Rightarrow (\mu(\varphi) \geq 0)$.

(a) If μ is a positive Radon measure on U , then the map $f: \mathcal{P}(A) \to \mathbf{R}$ defined by $f(X) = \mu(I_{\omega(X)})$ for $X \subset A$ is a measure on A . ($I_{\omega(X)}$ is the characteristic function of $\omega(X)$).

(b) Conversely, show that for any measure f on A there is a unique positive Radon measure μ on U such that $f(X) = \mu(I_{\omega(X)})$ for all $X \subset A$. (The term "extremal measure" will be justified later in section 37.)

4.5 <u>More on filters</u>. Give several further properties of filters analogous to properties of nets. (See Kelley [1] <u>General Topology</u>.)

§5 UNIFORM SPACES.

Two of the basic notions one can introduce in the category of metric spaces are uniform continuity and completeness. However, there is a need for these concepts in spaces which are not necessarily metrizable. The correct category in which to work is the category of uniform spaces.

There are two equivalent ways of defining a uniform space; one is by way of pseudo-metrics, the other by way of vicinities. We begin with the pseudo metric method.

5.1 DEFINITION. Let E be a set and $\{d_i\}$ a family of pseudo metrics on E. Let E_i denote E with the d_i topology and $f_i: E \to E_i$ the identity map. The initial topology on E is called the topology generated by the family $\{d_i\}$.

The family d_i is called saturated iff

$$(d_i, d_j \in \{d_i\}) \Rightarrow (\sup(d_i, d_j) \in \{d_i\})$$

where

$$\sup(d_i, d_j)(x,y) = \sup(d_i(x,y),\ d_j(x,y))$$

The reason we consider saturated families is the following: for $\{d_i\}$ saturated, the topology on E is generated by the $\{y \in E: d_i(x_0,y) < \varepsilon\}$ where x_0 ranges over E, $\varepsilon > 0$ and $d_i \in \{d_i\}$. If $\{d_i\}$ were not saturated this set would merely be a subbasis. The condition of being saturated is not restrictive, for if $\{d_i\}$ is not saturated, we can saturate it by adding all finite sup's of elements of $\{d_i\}$ (without affecting the topology). It is easy to see that the sup of two pseudo-metrics is again a pseudo-metric, but this does not hold for inf's.

5.2 EXAMPLE. Let E be a topological space and $\mathcal{C}(E,\mathbf{R})$ the space of continuous functions $f: E \to \mathbf{R}$. For each $K \subset E$ compact, let

$$p_K(f) = \sup\{|f(x)|: x \in K\}$$

Then the family of semi-norms p_K ($d_K(f,g) = p_K(f-g)$ is then a pseudo-metric) generates a topology on $\mathcal{C}(E,\mathbf{R})$ called the topology of uniform convergence on compacts.

This space should not be confused with $\mathcal{K}(E,\mathbf{R})$, the continuous functions with compact support. Here, for each compact $K \subset E$, let $\mathcal{K}_K(E,\mathbf{R})$ denote the space of continuous functions with support in K and with the uniform norm. The topology on $\mathcal{K}(E,\mathbf{R})$ is the final topology with respect to the inclusion maps

$$f_K\colon \mathcal{K}_K(E,\mathbf{R}) \to \mathcal{K}(E,\mathbf{R}) .$$

Thus $\mathcal{K}(E,\mathbf{R})$ is the inductive limit of the $\mathcal{K}_K(E,\mathbf{R})$. See §16 for more details. If E is compact, note that $\mathcal{C}(E,\mathbf{R})$ is just the usual space of continuous functions with the supremum norm (a Banach space).

Using pseudo-metrics, we can give meaning to the notions of uniform continuity and completeness just as in metric spaces. However, there is an intrinsic method, introduced by Weil, which is the following.

5.3 DEFINITION. A uniform structure, or uniformity on a set E is a set $\mathcal{U} \subset \mathcal{P}(E \times E)$ satisfying the following conditions:

(i) $\mathcal{U}$ is a filter on $E \times E$

(ii) $(X \in \mathcal{U}) \Rightarrow (\Delta \subset X)$, where

$\Delta = \{(x,x): x \in E\}$

(iii) $(X \in \mathcal{U}) \Rightarrow (X^{-1} \in \mathcal{U})$

(iv) $(X \in \mathcal{U}) \Rightarrow$ (there exists $Y \in \mathcal{U}$ such that $Y \circ Y \subset X$) .

(Recall that $X^{-1} = \{(x,y): (y,x) \in X\}$ and $Y \circ Y' = \{(x,y):$ there exists $z \in E$ such that $(x,z) \in Y$ and $(z,y) \in Y'\}$) . Elements of $\mathcal{U}$ are called <u>vicinities</u> or <u>entourages</u>.

A <u>uniform space</u> is a set together with a uniform structure on it.

Notice that these conditions are modelled on the axioms for a metric space. For example, (iv) is descendant from the triangle inequality.

5.4 <u>DEFINITION</u>. Let $(E, \mathcal{U})$ be a uniform space. The <u>uniform topology</u> on E is defined by: $V \subset E$ is open iff for each $x \in V$, there is a $U \in \mathcal{U}$ such that $U(x) \subset V$. (Recall that $U(x) = \{y \in E: (x,y) \in U\}$.)

To check that this is a topology is easy.

In fact the union of open sets is open, and to show that the intersection of two open sets is open, note that $U(x) \cap V(x) = (U \cap V)(x)$.

For the uniform topology it is important to note that the sets $U(x)$ (which are analogues of ε-discs) are not necessarily open. Rather, the $U(x)$ are a fundamental system of neighborhoods (in fact, the entire neighborhood system).

A uniform structure is inherited by subsets of a uniform space by restriction. As with topologies, the restriction is called the <u>relative uniformity</u>. Evidently the relative topology coincides with the topology of the relative uniformity.

5.5 <u>DEFINITION</u>. Given a uniform space $(E, \mathcal{U})$, we say that $\mathcal{U}$ is <u>compatible</u> with a topology $\mathfrak{T}$ on E iff $\mathfrak{T}$ coincides with the uniform topology. We say that a topology $\mathfrak{T}$ is <u>uniformizable</u> iff there is a uniform structure $\mathcal{U}$ compatible with $\mathfrak{T}$.

It is important to know that a topology can be uniformized in different ways. More on this will be given below.

5.6 PROPOSITION. *If $\{d_i\}$ is a saturated family of pseudo-metrics on E, then $\mathcal{U}$, consisting of sets of the form*

$$U_{d_i,\varepsilon} = \{(x,y) \in E \times E: d_i(x,y) < \varepsilon\}$$

is a filter base of a uniformity which is compatible with the topology generated by $\{d_i\}$. This uniformity is called the uniformity generated by $\{d_i\}$.

Proof. That $\mathcal{U}$ is a filter base follows because $\{d_i\}$ is saturated. To show $\mathcal{U}$ generates a uniformity, it suffices to check (ii), (iii) and (iv) of 5.3. But (ii) is clear as $d_i(x,x) = 0$ and (iii) follows by symmetry. For (iv) we note that

$$U_{d_i,\varepsilon/2} \circ U_{d_i,\varepsilon/2} \subset U_{d_i,\varepsilon}$$

which follows from the triangle inequality. For compatibility, just note that

$$U_{d,\varepsilon}(x) = \{y \in E: d(x,y) < \varepsilon\}$$

which is a neighborhood of x. □

Note that this uniformity just consists of certain neighborhoods of the diagonal in the product topology.

Perhaps not as clear is the fact that <u>all</u> uniform structures arise from a family of pseudo-metrics.

5.7 <u>THEOREM</u>. <u>Let</u> $(E, \mathcal{U})$ <u>be a uniform space</u>. <u>Then there exists a saturated family</u> $\{d_i\}$ <u>of pseudo-metrics on</u> E <u>which generates</u> $\mathcal{U}$ <u>in the sense of 5.6</u>.

<u>Proof</u>. For each $U \in \mathcal{U}$, symmetric (note that $U \cap U^{-1} \in \mathcal{U}$ is always symmetric), construct a sequence V_n in $\mathcal{U}$, with V_n symmetric, and such that $V_n \circ V_n \circ V_n \subset V_{n-1}$; and $V_0 = U$. Let $f(x,y) = 2^{-n}$ if $(x,y) \in V_{n-1} \setminus V_n$ and equal zero on $\cap \{V_n\}$. Define d_U by

$$d_U(x,y) = \inf\{\Sigma_{k=0}^{n} f(a_k, a_{k+1}) : a_0 = x,\ a_k \in E,\ a_n = y\}$$

Then d_U is a pseudo-metric and

$$V_n \subset \{(x,y) : d(x,y) < 2^{-n}\} \subset V_{n-1}$$

(These verifications we leave to the reader). If

we saturate this family $\{d_U\}$ it is obvious that $\mathcal{U}$ coincides with the uniformity generated by the $\{d_U\}$. □

5.8 DEFINITION. A uniform space is pseudo-metrizable [resp. metrizable] iff the uniformity (not just the uniform topology) is generated by a pseudo-metric [resp. metric].

We have the following very useful corollary of 5.7.

5.9 COROLLARY. A uniform space is pseudo-metrizable iff $\mathcal{U}$ has a countable filter base. (If the uniform topology is Hausdorff, pseudo-metrizable becomes metrizable).

Proof. If $\mathcal{U}$ is pseudo-metrizable then obviously $\mathcal{U}$ has a countable base, namely

$$U_n = \{(x,y): d(x,y) < 1/n\}$$

Conversely, if $\mathcal{U}$ has a countable base then we may choose the sequence V_n of 5.7 to be a base for $\mathcal{U}$. Hence the pseudo-metric associated to

V_n (constructed in 5.7) generates the uniformity. □

5.10 Remark. Let $\mathcal{U}$ be a uniformity on E with $\mathfrak{T}$ the uniform topology. If $\mathcal{U}$ is generated by the family $\{d_i\}$ then $\mathfrak{T}$ is Hausdorff iff

$$(x,y) \in E,\ x \neq y) \Rightarrow (\text{there exists } d_i \text{ such that } d_i(x,y) \neq 0)$$

This is also equivalent to $\cap\{U \in \mathcal{U}\} = \Delta$. If $\mathfrak{T}$ is not Hausdorff, introduce the equivalence relation $(xRy) \Longleftrightarrow (d_i(x,y) = 0$ for all $d_i)$. Then E/R has a natural structure of a Hausdorff uniform space generated by the inherited pseudometrics.

In this regard we give two warnings. First, quotient spaces do not in general inherit uniform structures. Second, this easy way to make a space Hausdorff does not work for general topological spaces. (For analogues in the category of topological spaces, see Sharpe et. al. [1].)

For metric spaces (E,d) and (F,ρ) recall that a map $f\colon E \to F$ is uniformly continuous iff for all $\varepsilon > 0$ there is a $\delta > 0$ such that

$$(d(x,y) < \delta) \Rightarrow (\rho(f(x), f(y)) < \varepsilon)$$

that is,

$$(f \times f)^{-1} (\{(a,b) \in F \times F: \rho(a,b) < \varepsilon\})$$
$$\supset \{(x,y): d(x,y) < \delta\}$$

This motivates the following definition.

5.11 DEFINITION. Let $(E, \mathcal{U})$ and $(F, \mathcal{V})$ be uniform spaces and $f: E \to F$ a mapping. We call f *uniformly continuous* iff

$$(U \in \mathcal{V}) \Rightarrow ((f \times f)^{-1} (U) \in \mathcal{U})$$

In terms of families of pseudo metrics we have:

5.12 PROPOSITION. *Let* $(E, \mathcal{U})$ *and* $(F, \mathcal{V})$ *be uniform spaces and let* $\{d_i\}$ *and* $\{\rho_\alpha\}$ *be saturated families of pseudo metrics generating* $\mathcal{U}$ *and* $\mathcal{V}$ *respectively.*

(i) *a map* $f: E \to F$ *is uniformly continuous iff for all* $\varepsilon > 0$ *and* ρ_α *there exists* $\delta > 0$ *and* d_i *such that*

$$(d_i(x,y) < \delta) \Rightarrow (\rho_\alpha(f(x), f(y)) < \varepsilon)$$

(ii) _If_ f _is uniformly continuous then it is continuous in the uniform topologies_.

(iii) _the composition of uniformly continuous maps is uniformly continuous_.

We leave the formal proof to the reader. Note that (i) follows from our remark preceding 5.11 and (ii) and (iii) may be easily deduced from (i), or proved directly.

Because of (iii) above and the obvious fact that the identity map is uniformly continuous, we have:

5.13 _PROPOSITION_. _The class of uniform spaces and uniformly continuous maps form a category_. _In this category we call an isomorphism a uniform isomorphism_.

In this category projective limits exist, but inductive limits do not. We leave the formal proof of this to the reader, and concentrate instead on the special case of initial and product uniformities.

5.14 <u>PROPOSITION</u>. <u>Let</u> E <u>be a set and</u> $\{(E_i, \mathcal{U}_i)\}$ <u>a family of uniform spaces. For mappings</u> $f_i : E \to E_i$, <u>there is a coarsest uniformity</u> $\mathcal{U}$ <u>on</u> E <u>such that each</u> f_i <u>is uniformly continuous</u>. (<u>Coarsest meaning the same as for filters</u>). <u>Furthermore the uniform topology on</u> E <u>is the initial topology relative to the uniform topologies on</u> $\{E_i\}$. <u>We call</u> $\mathcal{U}$ <u>the initial uniformity</u>.

<u>Proof</u>. Consider the family of all $(f_i \times f_i)^{-1}(U_i)$ for $U_i \in \mathcal{U}_i$. These form a subbase of a filter; that is, all finite intersections form a filter base. Let $\mathcal{U}$ be the corresponding filter on $E \times E$. The conditions of 5.3 are easily verified, so that $\mathcal{U}$ is a uniformity. Obviously $\mathcal{U}$ is the coarsest uniformity for which each f_i is uniformly continuous. To prove the last statement, observe that both topologies are generated by the system of neighborhoods

$$((f_i \times f_i)^{-1}(U_i))(x) = f_i^{-1}(U_i(f_i(x))) . \square$$

As usual we will call the initial uniformity of a product the <u>product uniformity</u>. Thus 5.14 implies that products exist in the category of uniform spaces. The analogue of 4.8 also holds. (We leave the proof to the reader).

Another consequence of 5.14 is:

5.15 COROLLARY. *Let* $\{\mathcal{U}_i\}$ *be a family of uniformities on a set* E . *There is a coarsest uniformity* $\mathcal{U}$ *finer than each* $\mathcal{U}_i$ *and its topology is that generated by the union of all the uniform topologies*.

For the proof we just apply 5.14 to the identity maps $f_i: E \to E_i$ where E_i is E with the uniformity $\mathcal{U}_i$. (In general $\bigcap \mathcal{U}_i$ is not a uniformity.)

In terms of pseudo-metrics, if in 5.14 $\mathcal{U}_i$ is generated by $\{d_{i\alpha}\}$ then $\mathcal{U}$ is generated by the family

$$\delta_{i\alpha}(x,y) = d_{i\alpha}(f_i(x), f_i(y)) .$$

In general we do not have final or even quotient uniformities. An example is projective space: the set of lines through the origin in a topological vector space with the quotient topology (although this fact is not obvious).

Next, we consider the topic of completeness of a uniform space.

5.16 DEFINITION. Let $(E, \mathcal{U})$ be a uniform space.

(i) A filter $\mathfrak{F}$ on E is called a Cauchy filter iff for all $U \in \mathcal{U}$ there exists $X \in \mathfrak{F}$ such that $X \times X \subset U$.

(ii) The uniform space is called complete iff every Cauchy filter converges, in the uniform topology.

(iii) A net x_α in E is called a Cauchy net iff for all $U \in \mathcal{U}$, there is an α_o such that $\beta, \alpha \geq \alpha_o$ implies $(x_\alpha, x_\beta) \in U$.

Some basic properties are:

5.17 PROPOSITION. *Let* $(E, \mathcal{U})$ *be a uniform space. Then*

(i) *every convergent filter [resp. net] is Cauchy*

(ii) *a Cauchy filter [resp. net] converges to each of its cluster points*

(iii) *a closed subspace of a complete uniform space* $(E, \mathcal{U})$ *is also complete,*

(iv) $(E, \mathcal{U})$ <u>is complete iff every Cauchy net converges</u>.

<u>Proof</u>. (i) Suppose $\mathfrak{F}$ converges to $x \in E$. For $U \in \mathcal{U}$, choose $V \in \mathcal{U}$ symmetric such that $V \circ V \subset U$. Now $V(x) \in \mathfrak{F}$ and $V(x) \times V(x) \subset U$, so $\mathfrak{F}$ is Cauchy. The proof for nets is similar.

(ii) Let x be a cluster point and $U \in \mathcal{U}$. Choose $V \in \mathcal{U}$ symmetric such that $V \circ V \subset U$ and X such that $X \times X \subset V$. But this means $X \subset U(x)$ since $X \cap V(x) \neq \emptyset$ and thus $X \subset V \circ V(x) \subset U(x)$.

(iii) If $A \subset E$ is closed and $\mathfrak{F}$ is a Cauchy filter on A, then $\mathfrak{F}$ is a filter base for a Cauchy filter on E, which converges to, say $a \in E$. Since A is closed, $a \in A$ and $\mathfrak{F}$ converges to a.

(iv) If $(E, \mathcal{U})$ is complete and x_α is is a Cauchy net then the filter of sections is a Cauchy filter, and so converges. This implies x_α converges. Conversely, let $\mathfrak{F}$ be a Cauchy filter and x_F, $F \in \mathfrak{F}$ a corresponding net. Then x_F is Cauchy with limit a. Then a is a cluster point of $\mathfrak{F}$ and hence by (ii), $\mathfrak{F}$ converges to a. □

Obviously this notion of completeness generalizes that for metric spaces. Thus, if $(E, \mathcal{U})$ is pseudo-metrizable, it is complete iff every Cauchy <u>sequence</u> converges. (<u>Proof</u>: From every Cauchy net one may select a subsequence which is also Cauchy).

The property of completeness is not a topological property, but depends on the choice of the uniformity. For example any metric topology has a complete compatible uniformity (see problem 5.4). For a space with no complete uniformity, see problem 6.2.

We now wish to show that any uniform space may be completed. To do this, we need a few preliminary results:

5.18 <u>PROPOSITION</u>. <u>Let</u> $\{(E_i, \mathcal{U}_i)\}$ <u>be a family of uniform spaces and</u> E <u>the product with the product uniformity</u>. <u>Then</u>

(i) <u>a filter</u> $\mathfrak{F}$ <u>on</u> E <u>is Cauchy iff</u> $\pi_i(\mathfrak{F})$ <u>is Cauchy on</u> E_i <u>for each</u> i ,

(ii) E <u>is complete iff each</u> E_i <u>is complete</u>.

Proof. (i) Suppose $\mathfrak{F}$ is Cauchy and $U \in \mathcal{U}_i$. Choose $X \in \mathfrak{F}$ such that $X \times X \subset (\pi_i \times \pi_i)^{-1}(U)$, or $\pi_i(X) \times \pi_i(X) \subset U$. Hence $\mathfrak{F}_i$ is Cauchy. Conversely suppose each $\mathfrak{F}_i$ is Cauchy. For each i, and $U \in \mathcal{U}_i$ there is an $X \in \mathfrak{F}$ such that $X \times X \subset (\pi_i \times \pi_i)^{-1}(U)$. Since these generate the uniformity, $\mathfrak{F}$ is Cauchy.

(ii) If each E_i is complete and $\mathfrak{F}$ is a Cauchy filter on E, then each $\mathfrak{F}_i$ is Cauchy by (i). Hence each $\mathfrak{F}_i$, and thus $\mathfrak{F}$ converge. Conversely suppose E is complete. Let x_α be a Cauchy net in E_i, and construct a net y_α in E whose i^{th} coordinate is x_α and the other coordinates are constant (axiom of choice). Hence y_α is Cauchy and so converges. Thus x_α converges. □

5.19 PROPOSITION. Every uniform space $(E,\mathcal{U})$ is uniformly isomorphic to a subspace of a product of pseudo-metric spaces. If E is Hausdorff, these pseudo-metric spaces may be chosen metric.

Proof. Let $\{d_i\}$ be a family of pseudo-metrics generating the uniformity and let E_i

denote E with the pseudo-metric d_i. Let F be the product of the E_i with the product uniformity. Define $\Psi: E \to F$ by $\Psi(x)_i = x$. It follows easily that Ψ is a uniform isomorphism onto its image.

In case E is Hausdorff we take F to be the product of E_i/R_i where xR_iy iff $d_i(x,y) = 0$, and define $\Phi: E \to F$ by $\Phi(x)_i = [x]_i$, where $[\cdot]_i$ denotes the equivalence class. Since E is Hausdorff, Φ is one to one and as above, is a uniform isomorphism onto its image. □

5.20 PROPOSITION. *Let* $(E,\mathcal{U})$ *be a uniform space and* $(F,\mathcal{V})$ *a complete Hausdorff uniform space. Let* $A \subset E$ *and* $f: A \to F$ *uniformly continuous. Then* f *has a unique extension to a uniformly continuous map* $g: c\ell(A) \to F$.

Proof. Uniqueness is obvious, for if $x \in c\ell(A)$ and $x_\alpha \to x$ with $x_\alpha \in A$, we must have $g(x) = \text{limit } f(x_\alpha)$ (and limits are unique as F is Hausdorff). For existence, note that if $x_\alpha \to x$ then $f(x_\alpha)$ is a Cauchy net in F as f is uniformly continuous. Let $g(x) = \text{limit } f(x_\alpha)$.

(g as defined is independent of the choice of the net $x_\alpha \to x$.) To see that g is uniformly continuous, we use 5.12 (i). Given $\varepsilon > 0$ and ρ_j, choose $\delta > 0$ and d_i such that $d_i(x,y) < \delta$ implies $\rho_j(f(x), f(y)) < \varepsilon$, for $x,y \in A$. Then also $\rho_j(g(x), g(y)) \leq \rho_j(g(x), g(x_\alpha)) + \rho_j(f(x_\alpha), f(y_\alpha)) + \rho_j(f(y_\alpha), g(y))$. When $x_\alpha \to x$, $y_\alpha \to y$, this gives the result. □

For an intrinsic, but less conceptual proof, see Kelley [1] <u>General Topology</u>, p. 195.

The completion theorem is the following:

5.21 <u>THEOREM</u>. <u>Every Hausdorff uniform space</u> E <u>is uniformly isomorphic to a dense subset of a complete Hausdorff uniform space</u> $\hat{E}$, <u>called the completion of</u> E. <u>Further</u>, $\hat{E}$ <u>is unique up to uniform isomorphism</u>.

<u>Proof</u>. By 5.19 we may embed E in a product F of metric spaces E_i. Let $\hat{E}_i$ be the completion of E_i and $\hat{F}$ the product. Thus we may embed E in $\hat{F}$, by f say. The desired completion is $c\ell(f(E))$. Uniqueness follows from 5.20. □

Finally we consider the relationship between completeness and compactness. The basic link is given by the notion of pre-compactness.

5.22 DEFINITION. Let $(E,\mathcal{U})$ be a uniform space. We say that E is *pre-compact* or *totally bounded* iff for each vicinity $U \in \mathcal{U}$, there is a finite set $x_1,\ldots,x_n \in E$ such that $E = \bigcup\{U(x_i)\}$.

5.23 PROPOSITION. *Let* $(E,\mathcal{U})$ *be a uniform space. Then the following are equivalent:*

(i) $(E,\mathcal{U})$ *is pre-compact*

(ii) *every ultrafilter on* E *is a Cauchy filter*

(iii) *for each* $U \in \mathcal{U}$ *there are subsets* $X_1,\ldots,X_n \subset E$ *with* $\bigcup X_i = E$ *and* $X_i \times X_i \subset U$.

(iv) *every net has a Cauchy subnet (a net* y_i, $i \in D'$ *is called a subnet of a net* $x_\alpha, \alpha \in D$ *iff there is an increasing mapping* $N: D' \to D$ *such that* $y_i = x_{N(i)}$ *and for each* $\alpha \in D$ *there is an* $i \in D'$ *such that* $(j \geq i) \Rightarrow (N(j) \geq \alpha))$.

Proof. (i) $\Rightarrow$ (iii). Given $U \in \mathcal{U}$, choose $V \in \mathcal{U}$ such that $V \circ V \subset U$. Now select $x_1,\ldots,x_n$

such that

$$E = \bigcup\{V(x_i) : i = 1,\ldots,n\}$$

and let $X_i = V(x_i)$.

(iii) $\Rightarrow$ (ii). Let $\mathfrak{F}$ be an ultrafilter on E and $U \in \mathcal{U}$. Choose $X_1,\ldots,X_n \subset E$ such that $E = \bigcup\{X_i\}$ and $X_i \times X_i \subset U$. By 4.10 (ii) (d), $X_i \in \mathfrak{F}$ for some i . Hence $\mathfrak{F}$ is a Cauchy filter.

(ii) $\Rightarrow$ (iv). Let x_α, $\alpha \in D$ be a net and $\mathfrak{F}$ an ultra-filter containing the filter of sections. Let $D' = \{(\alpha,F): \alpha \in D, F \in \mathfrak{F} , x_\alpha \in F\}$. Then D' is a directed set by: $(\alpha,F) \geq (\alpha',F')$ iff $\alpha \geq \alpha'$ and $F' \subset F$. Let $N: D' \to D$ be defined by $N(\alpha,F) = \alpha$ and consider the net $y_{\alpha,F} = x_\alpha$, which is a subnet of x_α . However, since $\mathfrak{F}$ is Cauchy and $y_{\alpha,F} \in F$, $y_{\alpha,F}$ is also Cauchy.

(iv) $\Rightarrow$ (i). Suppose E is not pre-compact. Then for some $U \in \mathcal{U}$, there is a sequence $x_1,x_2,\ldots$ such that $(m > n) \Rightarrow (x_m \notin \bigcup\{U(x_i): i = 1,\ldots,n\})$. But this means that the sequence x_n has no Cauchy subnet. □

This rather complicated definition of subnet is necessary as we saw in the proof (ii) ⇒ (iv). The non-equivalent version using subsets of D (containing arbitrarily large elements of D) will not suffice.

From this we deduce the following important result:

5.24 <u>THEOREM</u>. <u>Let</u> $(E,\mathcal{U})$ <u>be a uniform space</u>. <u>Then</u> E <u>is pre-compact iff its completion</u>, $\hat{E}$ <u>is compact</u>. <u>In particular</u> E <u>is compact iff</u> E <u>is complete and totally bounded</u>.

<u>Proof</u>. First we show that (E pre-compact) ⇒ ($\hat{E}$ pre-compact). Let $\{d_\alpha\}$ be a family of pseudo-metrics generating the uniformity of $\hat{E}$, and hence of E also, by restriction. Given $\epsilon > 0$ and α, there exist $X_1,\ldots,X_n$ of d_α diameter less than ϵ and such that $X_1 \cup \ldots \cup X_n = E$, since E is pre-compact. Then $c\ell(X_i)$ has d_α diameter $< \epsilon$ also, and $c\ell(X_1) \cup \ldots \cup c\ell(X_n) = \hat{E}$. Hence $\hat{E}$ is pre-compact.

Thus if E is pre-compact, every ultrafilter on $\hat{E}$ is Cauchy by 5.23 and hence converges, as

$\hat{E}$ is complete. This implies that $\hat{E}$ is compact, by 4.15.

Conversely, if $\hat{E}$ is compact then E is pre-compact by 5.23 (iii).

Finally, if E is pre-compact and complete, the above shows that E is compact. Conversely, if E is compact then every ultrafilter converges, so that E is pre-compact and complete, by 5.23, 5.17 (i) and 5.17 (ii) respectively. □

Finally we give a result on products of complete spaces which will be useful later.

5.25 PROPOSITION. *A countable product of complete (pseudo-) metric spaces is a complete (pseudo-) metric space.*

Proof. The product is complete by 5.18 and as the product is countable, the product is (pseudo-) metrizable by 5.9. □

PROBLEMS FOR §5

5.1 *Compact uniform spaces*

(i) Let $(E,\mathcal{U})$ and $(F,\mathcal{V})$ be uniform

spaces with $f: E \to F$. If f is continuous in the uniform topology it is uniformly continuous on compact subsets.

(ii) If a compact Hausdorff space is uniformizable (we will see this in §6) it has a unique uniformity.

5.2 Reduction to countable products.

(i) Let $\{\mathcal{U}_i\}$ be a family of uniformities on a set E with $\mathcal{U}$ the (least) upper bound. If $\mathcal{U}$ has a denumerable base of vicinities, then $\mathcal{U}$ is also the least upper bound of some finite or denumerable subfamily of $\{\mathcal{U}_i\}$.

(ii) Let E be a product of uniform spaces $\{E_i\}$, $i \in I$, and let $A \subset E$ be metrizable in the sense that the uniformity on A arises from a metric. Then there exists $J \subset I$ with J finite or denumerable such that the projection of A into $E_J = \prod_{i \in J} E_i$ is a uniform isomorphism. Prove the analogous result assuming instead that the topology on A has a countable base.

5.3 Complete uniformities.

(i) Let $(E,\mathcal{U})$ be a Hausdorff uniform space

If $\{E_\alpha\}$ is a family of subspaces of E, each of which has a complete uniformity, then $\cap\{E_\alpha\}$ has a complete uniformity, compatible with its topology.

(ii) If E has a complete uniformity, then any $\mathfrak{F}_\sigma$ in E has a complete uniformity compatible with its topology.

(iii) Any uniform space with a first countable completion has a complete uniformity. In particular, any metric space has a complete uniformity.

§6 CONSEQUENCES OF URYSOHN'S LEMMA

Urysohn's lemma ensures the existence of a rich supply of continuous functions on certain to-topological spaces. Using this tool, we prove some important embedding and metrization theorems. Another consequence of the lemma is the existence of partitions of unity, an extremely useful tool in itself.

We begin with the classical Urysohn and Tietze theorems. Since these theorems really belong to elementary topology we shall just sketch the proofs and leave the problem of establishing burdensome notations to the reader.

Recall that a topological space is <u>normal</u> iff any two disjoint closed sets have disjoint (open) neighborhoods. Obviously E is normal iff each closed set has a base of closed neighborhoods.

6.1 <u>THEOREM</u>. <u>Let</u> E <u>be a Hausdorff topological space</u>. <u>Then the following are equivalent</u>

(i) E <u>is normal</u>

(ii) <u>for any two closed, non-empty, disjoint sets</u> A, B, <u>there is a continuous function</u> $f: E \to [0,1]$ <u>such that</u> $f(A) = 0$ <u>and</u> $f(B) = 1$

(iii) *for any closed set* $A \subset E$ *and continuous function* $f: A \to [a,b]$, *there exists* $g: E \to [a,b]$ *which is continuous and is an extension of* f .

The equivalence (i) <=> (ii) is called *Urysohn's lemma* and (i) <=> (iii) is the *Tietze extension theorem*.

The idea of the proof is the following. First note that the implications (ii) ⇒ (i) and (iii) ⇒ (ii) are trivial. For the implication (i) ⇒ (ii), the idea is the following. By normality, there is an open set U such that

$$A \subset U \subset c\ell(U) \subset E \setminus B .$$

By repeating this argument we can find a whole family of such open sets $\{U_t: t \in D\}$ where D is a countable dense subset of $[0,1]$, and $t < s$ implies

$$A \subset U_t \subset c\ell(U_t) \subset U_s \subset c\ell(U_s) \subset E \setminus B .$$

The required function is defined by $f(x) = 0$ if x is in every U_t and $f(x) = \sup\{t: x \notin U_t\}$ otherwise. Continuity of f follows from the fact that

$$f^{-1}([0,a[) = \cup\{U_t:\ t < a\}$$

and

$$f^{-1}(]a,1]) = \cup\{E \quad c\ell(U_t):\ t > a\} .$$

Next we outline the method for proving that (ii) $\Rightarrow$ (iii). First note that in (ii), $[0,1]$ may be replaced by any interval $[a,b]$ (take $g = (b-a)f + a$), and that it suffices to prove (iii) for $[-1,1]$. Let $A_0 = \{x \in A:\ f(x) \leq -1/3\}$ $B_0 = \{x \in A:\ f(x) \geq 1/3\}$. Use (ii) to define a map $g_0\colon E \to [-1/3,\ 1/3]$ such that $g_0(A_0) = -1/3$ and $g_0(B_0) = 1/3$. Let $f_1 = f - g_0$ and repeat the procedure with f_1. Doing this, one obtains a sequence

$$f_n = f - (g_0 + g_1 + \dots + g_{n-1}) = f - s_n$$

such that $\|f_n\| \leq (2/3)^n$ (supremum norm), and $\|g_n\| \leq (1/3)(2/3)^n$. Hence s_n converges uniformly to a function g on E which agrees with f on A. □

With regard to this theorem, it is convenient to isolate the key property. Namely, a space E

is called _completely regular_ iff for each $a \in E$ and neighborhood U of a, there is a continuous map $f: E \to [0,1]$ such that $f(a) = 0$ and $f(E \setminus U) = 1$.

Thus, from 6.1, normal spaces are completely regular. Another useful fact is that metric spaces are normal.

A _cube_ is a space of the form $[0,1]^I$ for some set I (the product of I copies of $[0,1]$). If I is denumerably infinite, the cube is called the _Hilbert cube_. The Hilbert cube is metrizable with metric

$$d(x,y) = \sum_{n=1}^{\infty} 2^{-n} |x_n - y_n|$$

where x_n are the components of x ; $n = 1,2,\ldots$. (In fact the Hilbert cube is complete metrizable by 5.25).

6.2 _THEOREM_. _Let_ E _be a Hausdorff space_. _Then the following are equivalent_

(i) E _is completely regular_

(ii) E _may be embedded in a cube_

(iii) E _is uniformizable_ (_its topology may be derived from a uniformity_).

Proof. First (i) implies (ii). In fact, let $\mathfrak{B}$ denote the set of continuous maps $f: E \to [0,1]$ and let $F = [0,1]^{\mathfrak{B}}$, that is, $\Pi[0,1]_i$ for $i \in \mathfrak{B}$. Define

$$\Phi: E \to F ; \Phi(x)_f = f(x) .$$

Then Φ is continuous since, if $\pi_f: F \to [0,1]$ is a projection associated with $f \in \mathfrak{B}$, $\pi_f \circ \Phi = f$ is continuous. Also, Φ is one to one, since for every $x_1 \neq x_2$ there is an $f \in \mathfrak{B}$ with $f(x_1) \neq f(x_2)$. Finally, Φ is a homeomorphism onto its image; for if $U \subset E$ is open and $x \in U$ we can find a continuous $f: E \to \mathbb{R}$ so that $f(x) = 0$ and $f = 1$ outside U. Let $V = f^{-1}([0, 1/2[)$, an open neighborhood of x contained in U. Then $\Phi(V) = \Pi_f^{-1}([0, 1/2[) \cap \Phi(E)$. Thus Φ^{-1} is continuous.

Next we show (ii) implies (iii): $[0,1]$ is a uniform space and hence F, a product of uniform spaces, is uniform. Then $\Phi(E) \subset F$ is also uniform, and we have (iii).

Finally (iii) implies (i). Let $\{d_i\}$ be a saturated family of pseudo-metrics for the uniformity and U an open neighborhood of $x \in E$. There exists $\epsilon > 0$ and i so that $B_i(x,\epsilon) = \{y: d_i(x,y) < \epsilon\} \subset U$. Let $f(y) = (\inf\{d_i(x,y),\epsilon\})/\epsilon$. This is easily seen to fulfill the requirements. □

There is also an important refinement of this theorem which yields a basic metrization theorem:

6.3 THEOREM. *Let E be a Hausdorff space. Then the following are equivalent:*

(i) *E is regular with countable base*

(ii) *E may be embedded in the Hilbert cube*

(iii) *E is metrizable and separable.*

Proof. First note that the implications (ii) ⇒ (iii) and (iii) ⇒ (i) are trivial. For the implication (i) ⇒ (ii) it suffices, by the proof of 6.2, to construct a countable family $\mathcal{B}$ of continuous maps such that

(i) $\mathcal{B}$ separates points

and (ii) for each $x \in E$ and U a neighborhood of x, there exists a continuous map $f: E \to [0,1]$ such that $f \in \mathcal{B}$, $f(x) = 0$ and $f(E \setminus U) = 1$.

To do this, let $\{V_n\}$ be a countable base and $\{x_m\}$ a countable dense subset. For each pair

(x_m, V_n) with $x_m \in V_n$, choose $f_{n,m}: E \to [0,1]$ such that $f_{n,m} = 0$ on a neighborhood of x_m and $f_{n,m}(E \setminus V_n) = 1$. Then $\mathfrak{B} = \{f_{n,m}\}$ is the required family. (The $f_{n,m}$ exist since E is normal; see section 1). □

Finally in this section we consider partitions of unity.

6.4 DEFINITIONS. Let E be a topological space and $f: E \to \mathbb{R}$ a mapping. The support of f is the closure of $\{x \in E: f(x) \neq 0\}$ and is denoted $S(f)$.

A collection of subsets $\{A_i\}$ of E is called locally finite iff for each $x \in E$ there is a neighborhood U of x so that $U \cap A_i = \Phi$ except for finitely many indices i .

A partition of unity on E is a family of continuous mappings $\{\varphi_i: E \to [0,1] \subset \mathbb{R}\}$ such that

(i) $\{S(\varphi_i)\}$ is locally finite

and (ii) $\Sigma\{\varphi_i(x)\} = 1$ for each $x \in E$.

We say that a partition of unity $\{\varphi_i: i \in I\}$ is subordinate to a covering $\{A_j: j \in J\}$ of E iff $\{S(\varphi_i)\}$ is a refinement of $\{A_j\}$; that is, for every $i \in I$ there is a $j \in J$ so that $S(\varphi_i) \subset A_j$.

Note that by (i), the sum in (ii) has only a finite number of non-zero terms. Also, (ii) implies $E = \bigcup\{S(\varphi_i)\}$.

6.5 PROPOSITION. A Hausdorff space E is normal iff every locally finite open cover $\{V_i\}$ has a refinement $\{U_i\}$ (with the same index set) which is also an open cover and $U_i \subset c\ell(U_i) \subset V_i$ for each i.

Proof. First, suppose E is normal and $i \in I$ Well order I and define $J \subset I$ by:

$(j \in J) \iff$ (for all $i \leq j$ there exists $U_{i,j}$ such that $U_{i,j} \subset c\ell(U_{i,j}) \subset V_i$ and $\bigcup\{U_{i,j}\colon i \leq j\} \cup \{V_i\colon i > j\} = E$, and also $(i \leq k \leq j) \Rightarrow U_{i,k} = U_{i,j}$).

By the last condition, we may write U_i for $U_{i,j}$. We claim that $J = I$, which we shall prove by transfinite induction.

Fix $i_0 \in I$ and suppose $i \in J$ for all $i < i_0$. We must show that $i_0 \in J$. Now first we show that $E = \bigcup\{U_i\colon i < i_0\} \cup \{V_i\colon i \geq i_0\}$. Each

$x \in E$ lies in finitely many V_i, so each $x \in V_k$ for a largest index k. If $k \geq i_o$ then the assertion holds. If $k < i_o$ then $k \in J$, so that $E = \bigcup\{U_i: i \leq k\} \cup \{V_i: i > k\}$ and hence $x \in U_i$ for some $i \leq k$. Again the assertion holds. Secondly, let $A = E \setminus \bigcup\{U_i: i < i_o\} \cup \{V_i: i > i_o\}$. Then by the above, $A \subset V_{i_o}$, so by normality, there is an open set U_{i_o} such that $A \subset U_{i_o} \subset c\ell(U_{i_o}) \subset V_{i_o}$. With this choice it is clear that $i_o \in J$.

Consequently $J = I$ and the definition of J shows that $\{U_i\}$ is the desired covering.

Conversely, let A, B be disjoint closed sets and consider the open cover $\{E \setminus A, E \setminus B\}$. Hence there exist $U_i \subset c\ell(U_1) \subset E \setminus A$ and $U_2 \subset c\ell(U_2) \subset E \setminus B$ with $U_1 \cup U_2 = E$. Then $V_1 = E \setminus c\ell(U_1)$ and $V_2 = E \setminus c\ell(U_2)$ are the required neighborhoods of A and B. □

The basic theorem on partitions of unity is the following:

6.8 THEOREM. Let E be a normal space and $\{U_i\}$ a locally finite open cover of E. Then there exists a partition of unity $\{\varphi_i\}$ (with the same index set) subordinate to $\{U_i\}$.

Proof. By 6.5 there is an open cover V_i with $V_i \subset c\ell(V_i) \subset U_i$. Choose W_i such that $V_i \subset c\ell(V_i) \subset W_i \subset c\ell(W_i) \subset U_i$ and let $\Psi_i: E \to [0,1]$ have $\Psi_i(V_i) = 1$ and $\Psi_i(E\setminus W_i) = 0$. Then $\Psi = \sum_i \Psi_i$ is continuous by local finiteness. The required partition of unity is $\varphi_i = \Psi_i/\Psi$; $(\Psi(x) \geq 1)$. □

Concerning partitions of unity, the notion of paracompactness (due to Dieudonné) is useful:

6.7 DEFINITION. A topological space E is called paracompact iff E is Hausdorff and every open cover of E has a locally finite open refinement (which also covers E).

6.8 PROPOSITION. If E is paracompact then E is normal.

Proof. First, suppose $\{E_i\}$ is a locally finite family of closed sets. Then $\bigcup\{F_i\}$ is closed (the complement is seen at once to be open).

Let A be a closed set in E and $x \notin A$.

For each $y \in A$ take U_y, V_y disjoint neighborhoods of x and y. The cover $\{E \setminus A, V_y\}$ has a locally finite refinement, say W_i. The closures of the W_i not containing x form a closed neighborhood of A disjoint from x by our above remark. Hence E is regular. That is, for each A closed, $x \notin A$, A and x have disjoint neighborhoods.

The case for two disjoint closed sets proceeds in a similar manner. □

Actually, on a paracompact space we can do a bit better than on normal spaces, as follows:

6.9 THEOREM. *Let E be paracompact and $\{U_i\}$ be any open covering of E. Then there is a partition of unity $\{\varphi_i\}$ (with the same index) subordinate to $\{U_i\}$.*

Proof. By 6.6 and 6.8 it suffices to show that there is a locally finite refinement $\{V_i\}$ of $\{U_i\}$ (with the same index set I). In fact let $\{W_j\}$ be a locally finite refinement of U_i, $(j \in J)$. For each j, choose (axiom of choice) an $i = i(j)$ so that $W_j \subset U_i$. Now define

$$V_{i_o} = \bigcup\{W_j : i(j) = i_o\} .$$

This is the required refinement. □

Notice that $V_i \subset U_i$, although V_i may be empty.

Basic properties of paracompact spaces are as follows:

6.10 PROPOSITION. (i) Every closed subset of a paracompact space is paracompact.

(ii) If E is compact and F is paracompact then $E \times F$ is paracompact.

(iii) The following spaces are paracompact

(a) compact Hausdorff spaces

(b) regular spaces with countable base

(c) locally compact Hausdorff spaces with countable base

(d) metrizable spaces.

(iv) If E is locally compact Hausdorff and connected, then E is paracompact iff E is σ-compact (the union of a countable collection of compact sets).

(v) If E is locally metrizable, then E is metrizable $\Longleftrightarrow$ it is paracompact.

We shall outline the proof of (iii) c. We can write $E = \cup\{U_n : n = 1,2,\dots\}$ where U_n are open, $c\ell(U_n)$ is compact and $U_n \subset c\ell(U_n) \subset U_{n+1}$. Let $\{V_\alpha\}$ be an open covering of E and $K_n = c\ell(U_n)\setminus U_{n-1}$. Since K_n is compact, it is covered by a finite number of open sets contained in some $V_\alpha \cap U_{n+1}$ and disjoint from $c\ell(U_{n-2})$. The union of these finite collections is the desired refinement of $\{V_\alpha\}$.

For the rest of the proofs the essential idea is the same as the above. For details, see Kelley [1] General Topology pp. 156-173 and Mokobodzki [1].

For some purposes one must obtain C^∞ partitions of unity. This may be done on any finite dimensional manifold and on certain infinite dimensional ones. See Lang, Introduction to Differentiable Manifolds, pp. 25-30. We shall not need these in our work.

PROBLEMS FOR §6

6.1 Locally uniformizable and metrizable spaces

(i) Let E be a regular topological space. Show that if every point of E has a uniformizable neighborhood, then E is uniformizable.

(ii) Let E be a compact space every point of which has a metrizable neighborhood. If E is Hausdorff, then E is metrizable.

Extend this result to locally compact, σ-compact spaces. Give a counterexample in case E is not Hausdorff.

6.2 The ordinal space

Let E be a non-denumerable, well ordered set such that for each $a \in E$, $\{x \in E: x \leq a\}$ is countable. Give E the topology generated by the base of open sets of the form $\{x \in E: a < x < b\}$ for $a, b \in E$, called the _order topology_.

(i) Show that E is locally metrizable, but not metrizable.

(ii) Show that any continuous map $f: E \to \mathbb{R}$ is eventually constant, that is constant on some segment $\{x \in E: x > a\}$. Prove an analogous result for $E \times E$.

(iii) Show that E has a unique uniformity.

(iv) Show that E with that uniformity is

not complete.

Note: E is not paracompact. See problem 6.3(v). This also follows from (i) and 6.10(v).

6.3 Paracompact spaces and complete uniformities

Let E be a Hausdorff space.

(i) An open cover $\{V_i\}$ of E is called even iff there is a neighborhood U of the diagonal in $E \times E$ such that for each $x \in E$, there exists V_i such that $U(x) \subset V_i$. Show that, if E is paracompact, then any open cover is even.

(ii) Let $\{V_i\}$ be a covering of E. Define the star of $B \subset E$ by $St(B, \{V_i\}) = \cup\{V_i : B \cap V_i \neq \phi\}$. A covering $\{U_\alpha\}$ is called a barycentric refinement of $\{V_i\}$ iff $\{St(\{x\}, U_\alpha)\}$ refines $\{V_i\}$. If E is paracompact, show that every open covering has an open barycentric refinement.

Note: The conditions in (i) and (ii) are equivalent to paracompactness. (See Kelley or Dugundji).

Let E be a paracompact space.

(iii) Show that the family of all neighborhoods of Δ in $E \times E$ is a uniformity for E.

(iv) Show that E is complete in the

uniformity in (i).

(v) Deduce that the ordinal space E in problem 6.2 is not paracompact.

§7 BAIRE SPACES

Around 1899, Baire introduced (on the real line) the notion of category. The idea was to find a topological counterpart to the concept of "almost everywhere". The concept of Baire category has proven to be one of the basic tools of analysis and is especially important in infinite dimensional spaces.

This section begins with some elementary consequences of the baire property, such as the Banach-Steinhaus theorem. We also include an application to the Hausdorff metric. The Baire category theorem is proven using the elegant concept of α-favorable spaces, which will also be used in later work. We conclude the section with some examples illustrating the qualitative (that is "almost everywhere") aspects of Baire spaces.

7.1 DEFINITION. Let E be a topological space and $X \subset E$. We call X nowhere dense iff $c\ell(X)$ has empty interior. (Sometimes nowhere dense sets are called meager.) A countable union of nowhere dense sets is called a set of the first category (or <u>rare</u>). Otherwise, a set is of the <u>second category</u> (all relative to E).

The space E is called a <u>Baire space</u> iff for any set $X \subset E$ of first category, $E \setminus X$ is everywhere dense in E.

In a Baire space E, the complement of any subset of the first category is called a <u>residual</u> of E.

A property P on E is said to hold <u>generically</u> or <u>almost everywhere</u> iff $\{x \in E: P(x)$ is true$\}$ is a residual set.

Notice that E being a Baire space means more than E is of second category.

Recall that an $\mathfrak{F}_\sigma$ set is a countable union of closed sets and that a $\mathcal{G}_\delta$ is a countable intersection of open sets.

7.2 <u>PROPOSITION</u>. <u>Let</u> E <u>be a topological space</u>.

(i) <u>The following are equivalent</u>:

(a) E <u>is a Baire space</u>

(b) <u>any countable intersection of open dense sets is dense</u>

(c) <u>the complement of every first cate-</u>

gory $\mathcal{F}_\sigma$ *is a dense* $\mathcal{G}_\delta$.

(ii) *if* E *is a Baire space then* E *is of second category*

(iii) *an open subset of a Baire space is Baire*

(iv) *if* E *is locally a Baire space* (*each* $x \in E$ *has a neighborhood* U *such that* $c\ell(U)$ *is a Baire space*), *then* E *is a Baire space*.

Proof. (i) (a) $\Rightarrow$ (b). Let $A = \bigcap\{U_n\}$ where U_n is open and dense. Then $X = \bigcup(E \setminus U_n)$ is of first category so that $A = E \setminus X$ is dense in E .

(b) $\Rightarrow$ (a). Suppose $X = \bigcup\{F_n\}$ is of first category. Let $A = \bigcap\{E \setminus c\ell(F_n)\}$. If F_n is nowhere dense, then $E \setminus c\ell(F_n)$ is open and dense so that A is dense. But $A \subset E \setminus X$ so that $E \setminus X$ is dense.

(a) $\Rightarrow$ (c) obvious.

(c) $\Rightarrow$ (b) proceeds as in (a) $\Rightarrow$ (b) .

(ii) If E were of first category, $\emptyset = E \setminus E$ would be dense in E .

(iii) Follows from the definition and the following two facts. First, a nowhere dense set in an open set is nowhere dense in the whole space.

Second, a subset dense in E is also dense in any open subset.

(iv) We use (i) (b). Let $A = \bigcap\{U_n\}$ where U_n is open and dense in E. Let U be a neighborhood of $x \in E$ as in the hypotheses. Then $A \cap c\ell(U) = \bigcap\{U_n \cap c\ell(U)\}$ is residual in $c\ell(U)$, as $U_n \cap c\ell(U)$ is dense in $c\ell(U)$. Therefore $c\ell(A) \cap c\ell(U) = c\ell(U)$, or $c\ell(U) \subset c\ell(A)$. Hence $c\ell(A) = E$. □

Note that any countable intersection of open dense sets is residual (an open dense set has a nowhere dense complement).

There is another simple but rather important characterization of Baire spaces, which is the following:

7.3 THEOREM. *Let* E *be a topological space. Then* (E *is Baire*) $\iff$ (*for any countable family of closed sets* $\{F_n\}$ *satisfying* $E = \bigcup\{F_n\}$ *we have* $\bigcup\{\text{int}(F_n)\}$ *is dense in* E).

Proof. ($\Rightarrow$). Let $G_n = F_n \setminus \text{int}(F_n)$ so that $X = \bigcup\{G_n\}$ is of first category. Hence $E \setminus X$ is dense in E. But $E \setminus X \subset \bigcup \text{int}(F_n)$.

($\Leftarrow$). Suppose E is not Baire. Then there exist open dense sets U_n such that $A = \bigcap\{U_n\}$ is not dense. Then $E = c\ell(A) \cup (\bigcup\{E \setminus U_n\})$, so we let $F_0 = c\ell(A)$, $F_n = E \setminus U_n$ and note that

$$\bigcup\{\text{int}(F_n)\} = \text{int}(c\ell(A)) \cup (\bigcup\{\text{int } F_n\})$$
$$= \text{int}(c\ell(A))$$

which is not dense, a contradiction. □

The following is a typical and important application of the notion of a Baire Space.

7.4 THEOREM (Banach-Steinhaus). *Let* E *and* F *be normed spaces with* E *a Baire space (for example,* E *is a Banach space; see* 7.12 *and* 7.13). *Let* $f_i: E \to F$ *be a family of continuous linear maps such that for each* $x \in E$, $\{\|f_i(x)\|\}$ *is bounded. Then* $\{\|f_i\|\}$ *is bounded.* (*This is often called the uniform boundedness theorem.*)

Proof. Let $g(x) = \sup\{\|f_i(x)\|\}$ and note that $E_\lambda = \{x: g(x) \leq \lambda\} = \bigcap_i\{x: |f_i(x)| \leq \lambda\}$ is closed. By 7.3, $\bigcup\{\text{int}(E_n)\}$ is dense in E . In particular there is an ϵ-ball B about some x_0 and a constant M such that for $x \in B$, $g(x) \leq M$. But

$$\|f_i\| = \sup\{\|f_i(x - x_o)\|/\|x - x_o\| : \|x - x_o\| = \epsilon\}$$
$$\leq [M + g(x_o)]/\epsilon . \square$$

This theorem extends easily to multilinear mappings, a result which is useful in some applications. We leave the statement and proof to the reader.

Theorem 7.4 has many important applications in functional analysis. For example, it may be used to prove the existence of continuous functions whose Fourier series diverges at one point. (See Horvath, [1] Topological Vector Spaces and Distributions, pp. 64-66.)

7.5 THEOREM. *Suppose* E *is a Baire space,* F *is a separable metric space and* $f : E \to F$ *is of first Baire class (that is, the inverse image of any open set is an* $\mathfrak{F}_\sigma$.) *Then* f *is continuous at every point of a dense* $\mathcal{G}_\delta$ *set*.

Proof. For each $n = 1,2,\ldots$, cover F with a countable family of open sets $\{\omega_m^n : m = 1,2,\ldots\}$ of diameter $< 1/n$. Let $f^{-1}(\omega_m^n) = \cup\{F_{m,k}^n :$

$k = 1,2,\ldots\}$ where $F^n_{m,k}$ is closed. Now $E = \bigcup\{F^n_{m,k}: k = 1,2,\ldots, m = 1,2,\ldots\}$, and hence by 7.3, $\Omega_n = \bigcup\{\mathrm{int}(F^n_{m,k}): k = 1,2,\ldots, m = 1,2,\ldots\}$ is an open set dense in E. The oscillation of f at points of Ω_n is less than $1/n$, so that f is continuous at the points of $\Omega = \bigcap\{\Omega_n: n = 1,2,\ldots\}$ which is dense by 7.2 (b). □

7.6 COROLLARY. *Let E be a Baire space in which every open set is an* $\mathcal{F}_\sigma$. *Then a lower [resp. upper] semi-continuous real valued function on E is continuous at the points of a dense* $\mathcal{G}_\delta$ *set.*

Proof. We claim that if g is lower semi-continuous, then g is of first Baire class, so that 7.5 will apply. This assertion follows at once from the fact that for $a, b \in \mathbb{R}$, $a < b$ we have:

$$g^{-1}(]a,b[) = \bigcup\{g^{-1}(]-\infty, b-1/n])\cap g^{-1}(]a,\infty[); n = 1,2,\ldots\}.$$

The assertion for $f: E \to \mathbb{R}$ upper semi-continuous is obtained by applying the above result to $-f$. □

Other related results may be found in the

problems for this section.

We now consider an application with a more topological flavor. (The reader may proceed to 7.11 without loss of continuity.)

7.7 DEFINITION. Let (E,d) be a metric space. For $a \in E$ and $B \subset E$, $B \neq \emptyset$, let

$$d(a,B) = \inf\{d(a,b): b \in B\}$$

and for $A \subset E$, $A \neq \emptyset$, let

$$\beta(A,B) = \sup\{d(a,B): a \in A\}$$

and
$$\rho(A,B) = \sup\{\beta(A,B), \beta(B,A)\} .$$

We also put $\rho(\emptyset.\emptyset) = 0$ and $\rho(\emptyset,A) = \infty$. We call ρ the Hausdorff metric.

Further, for topological spaces E and F and a mapping $f: F \to \mathcal{P}(E)$ is called upper [resp. lower] semi-continuous iff for each U open [resp. closed] in E, $\{x \in F: f(x) \in U\}$ is open [resp. closed] in F. (This does not depend on any topology for $\mathcal{P}(E)$).

The next proposition will be left as a verification for the reader.

7.8 PROPOSITION. *Let* (E,d) *be a metric space. Then* (i) *the Hausdorff metric* ρ *is a (generalized) pseudo-metric on* $\mathcal{P}(E)$ *and is a (generalized) metric on* $\mathfrak{F}$, *the closed subsets of* E .

(ii) *if* F *is a topological space and* $f: F \to \mathcal{P}(E)$ *is a mapping then* f *is continuous for the Hausdorff metric* ρ *iff for all* $x \in F$ *and* $\varepsilon > 0$ *there is a neighborhood* U *of* x *such that*: $(y \in U) \Rightarrow$ (*for all* $a \in f(y)$, *there exists* $b \in f(x)$ *so that* $d(a,b) < \varepsilon$ *and for all* $c \in f(x)$ *there exists* $d \in f(y)$ *such that* $d(c,d) < \varepsilon$.)

The Hausdorff metric topology actually depends on the metric d on E , and not merely on the topology of E . However continuity of a map $f: F \to \mathcal{P}(E)$ often does not depend on d , as the next proposition shows.

7.9 PROPOSITION. *Let* E *be a topological space and* (K,d) *a compact metric space. Let* $\mathfrak{F}$ *denote the closed subsets of* K *with the Hausdorff metric. A mapping* $f: E \to \mathfrak{F}$ *is continuous* [*resp. upper, lower semi-continuous*] iff for all $\varphi \in \mathcal{C}(K,\mathbb{R})$, *the map* $x \mapsto A(\varphi,f(x))$ *is continuous*, [*resp.*

upper, lower semi-continuous] *where* $A(\varphi,F) = \sup\{\varphi(x): x \in F\}$. *Also,* f *is continuous iff* f *is both upper and lower semi-continuous*.

Proof. Suppose f is continuous. Given $\epsilon > 0$, choose $\eta > 0$ so that $d(a,b) < \eta$ implies $|\varphi(a) - \varphi(b)| < \epsilon$. By 7.8, find a neighborhood U of x_0 so that the conclusion of 7.8 (ii) holds with η. Suppose $A(\varphi, f(x_0)) = \varphi(a)$ and $A(\varphi, f(x)) = \varphi(c)$ for $a \in f(x_0)$, $c \in f(x)$, and $x \in U$. Now there exists, by 7.8 (ii), $b \in f(x)$ so that $d(a,b) < \eta$ and $d \in f(x_0)$ so that $d(c,d) < \eta$. Hence we have

$$\varphi(b) \leq \varphi(c) \ ; \ |\varphi(b) - \varphi(a)| < \epsilon \ ,$$

$$\varphi(d) \leq \varphi(a) \ ; \ |\varphi(d) - \varphi(c)| < \epsilon$$

and thus, $\varphi(a) - \epsilon \leq \varphi(b) \leq \varphi(c) \leq \varphi(d) + \epsilon < \varphi(a) + \epsilon$ which implies $|\varphi(c) - \varphi(a)| < \epsilon$. The proof of the converse and the semi-continuous statements are left to the reader. □

7.10 THEOREM. *Let* E *be a Baire space and* K *a compact metric space. Suppose* f *is lower semi-continuous;* $f: E \to \mathfrak{F}$. *Then there exists a dense*

$\mathscr{G}_\delta$ <u>in</u> E <u>at each point of which</u> f <u>is continuous</u>.

<u>Proof</u>. There exists a countable dense subset $\{\varphi_n\} \in \mathcal{C}(K,\mathbb{R})$. This may be seen as follows: Let $\{a_n\}$ be a countable dense set in K and $f_n(x) = d(x,a_n)$. Consider the algebra generated by these f_n with rational coefficients. It separates points as the $\{a_n\}$ are dense, and for $x \in E$ contains a function non-vanishing at x. Hence by the Stone Weierstrass theorem this algebra is dense.

Hence by 7.6 there is a dense $\mathscr{G}_\delta$, say G_n on which $x \mapsto A(\varphi_n, f(x))$ is continuous. Let $G = \bigcap\{G_n\}$, a dense $\mathscr{G}_\delta$ on which each of these maps is continuous. Since $\{\varphi_n\}$ are dense, and $\|A(\varphi,F) - A(\Psi,F)\| \leq \|\varphi - \Psi\|$, we have continuity of the mapping $x \mapsto A(\varphi,f(x))$ for all φ at points $x \in G$. □

Next we turn to techniques for deciding when a given space is Baire, stemming from game theory.

Before giving precise definitions, we give a heuristic discussion. Let E be a topological space and α,β be two "players" with β the first to move. Two games are:

(i) Each player chooses a non-empty open set V in E lying in the opponent's previously chosen open set.

(ii) Same as (i), except a point is chosen by β in V and α's open set must contain the point.

The space E is called α-favorable [resp. strongly α-favorable] if α has a winning tactic in game (i) (resp. (ii)), that is, if α can choose sets V_n so that $\bigcap\{V_n\} \neq \emptyset$ no matter what β chooses (in which case α "wins"). Usually strategies depend on all previous moves, but here they depend only on the previous move.

More precisely, we make the definition:

7.11 DEFINITION. Let E be a topological space. We say that E is α-favorable iff there exists a map (called the winning tactic) $f\colon \mathfrak{J}^* \to \mathfrak{J}^*$; such that $f(U) \subset U$, where $\mathfrak{J}^*$ denotes the collection of non-empty open sets of E, and such that for any sequence $V_1, V_3, \ldots$ defined inductively so that

$$V_1 \supset V_2 = f(V_1) \supset V_3 \supset f(V_3) \supset \ldots$$

we have $\bigcap\{V_n\} \neq \emptyset$. Similarly, a space E is called <u>strongly</u> α-favorable iff there exists a map $f: \mathcal{P} \to \mathcal{J}^*$ where $\mathcal{P} = \{(U,x): U \in \mathcal{J}^* \text{ and } x \in U\}$ and $\mathcal{J}^*$ is as above, such that $x \in f(U,x)$, $f(U,x) \subset U$ and such that for any sequence of pairs (V_1, x_1) , $(V_3, x_3), \ldots$ of $\mathcal{P}$ defined inductively such that

$$V_1 \supset V_2 = (V_1,x_1) \supset V_3 \supset f(V_3,x_3) \supset \ldots$$

we have $\bigcap\{V_n\} \neq \emptyset$.

In this game for strongly α-favorable, notice that β chooses the point about which α must place his open set but α does not make such a choice. In fact, if α were allowed to choose a point about which β must place his open set the game would be trivial. Evidently if E is strongly α-favorable then E is α-favorable.

For (strongly) α-favorable spaces we have two fundamental theorems which are the following:

7.12 <u>THEOREM</u>. <u>The following spaces are</u> α-<u>favorable</u> [<u>resp. strongly</u> α-<u>favorable</u>]:

(i) <u>complete metric spaces</u>

(ii) locally compact Hausdorff spaces

(iii) open subsets of α-favorable spaces [resp. strongly α-favorable spaces]

(iv) products of α-favorable spaces [resp. strongly α-favorable spaces].

7.13 THEOREM. *If E is α-favorable then E is Baire.*

Proof of 7.12. We shall give the proof in the α-favorable case and leave the (similar) proof of the strongly α-favorable case to the reader.

(i) For f define $f(U)$ such that diameter $f(U) \leq (\inf\{1, \text{diameter } U\})/2$ and $f(U) \subset cl(f(U)) \subset U$. Then given V_i, $i = 1,2,\ldots,$ and $x_n \in V_n$, x_n is a Cauchy sequence and its limit lies in $\cap\{V_n\}$.

(ii) Here choose $f(U)$ with $cl(f(U))$ compact and $cl(f(U)) \subset U$. Then

$$\cap V_i \neq \emptyset$$

since a nest of closed, non-empty sets in a compact space has non-empty intersection.

(iii) is obvious; use the same tactic as is

used for E .

(iv) suppose each E_i, $i \in I$ is α-favorable with a winning strategy $f_i: \mathcal{J}_i^* \to \mathcal{J}_i^*$. Define f as follows: given $U \subset \Pi E_i$, find $U_i \subset E_i$ such that $\Pi U_i \subset U$ (with all but a finite number of $U_i = E_i$) and let $f(U) = \Pi f_i(U_i)$ (we set all but a finite number of factors equal to E_i) . Suppose we have

$$V_1 \supset V_2 = f(V_1) \supset V_3 \supset f(V_3) \supset V_5 \supset f(V_5) \supset \dots$$

in ΠE_i . Now let $f(V_{2n+1}) = \Pi V_{2n+1}^i$ so that

$$\cap\{f(V_{2n+1})\} = \prod_i \{\cap\{V_{2n+1}^i : n = 1,2,\dots \}\} \neq \emptyset$$

since each $\cap\{V_{2n+1}^i : n = 1,2,\dots \} \neq \emptyset$.

Proof of 7.13. Suppose E is not Baire. Then by 7.2 (i), (a) $\Longleftrightarrow$ (c), there are closed nowhere dense sets F_n such that there is a non empty open set $V \subset \mathrm{int}(\cup\{F_n\})$. Let $V_1 = V$ and in general, let $V_{2n+1} = V_{2n} \cap (E \setminus F_n)$. There is no f giving a winning strategy since $V \cap (E \; (\cup\{F_n\})) = \emptyset$. □

From this we deduce for example that complete metric spaces and locally compact Hausdorff spaces are Baire. (Usually called the Baire Category Theorem). But 7.12 (iv) gives much more. Namely arbitrary products of these spaces are also Baire. For example, $\mathbb{R}^I$ is Baire for any set I. This is important for it is unknown in general if the product of metric Baire spaces is Baire.

Finally in this section we illustrate some interesting generic (almost everywhere) properties.

7.14 EXAMPLE. We shall prove that almost every $f \in \mathcal{C}([0,1], \mathbb{R})$ has no derivative at any point $x_0 \in [0,1]$; that is, the collection of such f is a residual (in the supremum (uniform) topology). (Weierstrass was the first to construct an example of such an f; see Gelbaum and Olmsted, [1] Counterexamples in Analysis, p. 38.)

Proof. Given $\epsilon > 0$, let

$$E_\epsilon = \{f \in \mathcal{C}([0,1], \mathbb{R}) \text{: for all } x_0 ,$$

$$\text{diameter } \{[f(x_0 + h) - h(x_0)]/h\text{: } \epsilon/2 \leq |h| \leq \epsilon\} > 1\}$$

Then E_ϵ is dense in $\mathcal{C}([0,1]; \mathbb{R})$. (Hint: for any polynomial $p \in \mathcal{C}([0,1], \mathbb{R})$, $p(x) + \eta \sin(kx) \in E_\epsilon$ for η sufficiently small and ηk sufficiently large.)

Also, E_ε is open as is seen at once from the definition. Therefore $E = \bigcap\{E_{1/n}\}$ is a residual set (a dense $\mathcal{G}_\delta$). However if $f \in E$ then f cannot have a derivative at any point. □

7.15 EXAMPLE. Let E be the Banach space of continuous maps $f: \mathbb{R}^n \to \mathbb{R}^n$ with support in some fixed compact K and with the supremum norm.

Consider the ordinary differential equation $d\varphi(t)/dt = f(\varphi(t))$; $\varphi: \mathbb{R} \to K \subset \mathbb{R}^n$, of class C^1 being the solution. It is well known that for each $(t_0, x_0) \in \mathbb{R} \times \mathbb{R}^n$, there exists a solution $\varphi: \mathbb{R} \to \mathbb{R}^n$ such that $\varphi(t_0) = x_0$, but that φ need not be unique. (See Coddington and Levinson, [1] Theory of Ordinary Differential Equations). However, if f is of class C^1 (or Lipschitz), the solutions are unique.

We shall show that generically (almost every) $f \in E$ has unique solutions through each point (t_0, x_0).

Proof. We say that f has ε-unique solutions for a time n iff for any two solutions φ, Ψ for f

through a common point, we have

$$\|\varphi(t) - \Psi(t)\| < \epsilon \text{ for all } t \in [-n,n].$$

For $h \in E$ of class C^1 we claim that for any $\epsilon > 0$ there is an $\eta > 0$ such that $\|f\| < \eta$, $f \in E$, implies $g = f + h$ has ϵ-unique solutions for a time n. To prove this, suppose for some $\epsilon > 0$ there is a sequence f_n with $\|f_n\| \to 0$ and such that there exist solutions φ_n, Ψ_n, for $g_n = f_n + h$, through a common point with $\|\varphi_n(t_n) - \Psi_n(t_n)\| \geq \epsilon$ for some $t_n \in [-n,n]$. Note that φ is a solution for g with $\varphi(t_0) = x_0$ iff

$$\varphi(t) = \int_{t_0}^{t} g(\varphi(\tau))d\tau + \varphi(t_0)$$

so that $\{\varphi_n\}$ and $\{\Psi_n\}$ are equicontinuous families from $[-n,n]$ to K. Therefore by the Arzela-Ascoli theorem, subsequences converge uniformly to φ and Ψ. Evidently φ and Ψ are distinct and are solutions for h; contradiction.

Now let $E_{\epsilon,n}$ denote those $f \in E$ which have ϵ-unique solutions for a time n. The above shows that $E_{\epsilon,n}$ is open and clearly E_ϵ is dense as it contains the f of class C^1. Then

$\cap\{E_{1/n,m}: n=1,2,\ldots,\ m=1,2,\ldots\}$ is the required residual $\mathcal{G}_\delta$. □

This theorem has an immediate generalization to continuous vectorfields X on a manifold. (See Abraham, [1] Foundations of Mechanics for terminology). There is also a more complicated generalization to vectorfields $X \in L_\infty$ (the φ here are not differentiable, so a new meaning of "solution" is needed). For machinery needed to make this precise, see Marsden [1]. This is an instance where the topology is crucial for a property to be generic. Indeed, uniqueness is generic for $X \in L_\infty$ but not for $X \in L_p$, $1 \le p < \infty$.

For a discussion of other generic properties of vectorfields, see Abraham, Foundations of Mechanics ch. V.

Our last example is a topological one.

7.16 EXAMPLE (Snake-Like Continua).) References for this example are R.H. Bing [1], [2], B. Knaster [1] and J. Hocking and G. Young, Topology, pp. 139-145.

A set $X \subset \mathbb{R}^2$ is called a continuum iff X

is compact and connected. Let $\mathcal{C}$ denote the set of all continua (in $\mathbb{R}^2$) equipped with the Hausdorff metric. It is clear that $\mathcal{C}$ is a metric space. Also, $\mathcal{C}$ is complete and hence is a Baire space (exercise for the reader).

A <u>Jordan domain</u> is the interior of a Jordan curve (a simple, closed continuous curve), in $\mathbb{R}^2$. A continuum X is called ϵ-<u>snake-like</u> iff it is contained in a Jordan domain D which can be covered by Jordan domains $D_1, \ldots, D_n$, such that diameter $(D_p) < \epsilon$, $D_p \cap D_{p+1} \neq \emptyset$, and $D_p \cap D_{p+k} = \emptyset$ if $k > 1$. A continuum X is called <u>snake-like</u> iff it is ϵ-snake-like for any $\epsilon > 0$.

The following illustrate snake-like continua.

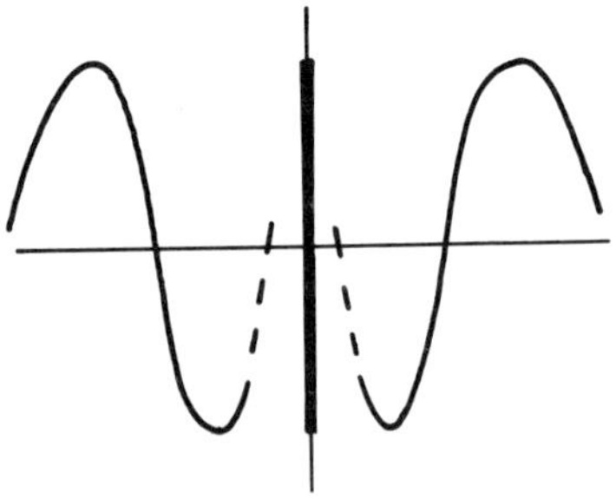

closure of graph of $\sin(1/|x|)$

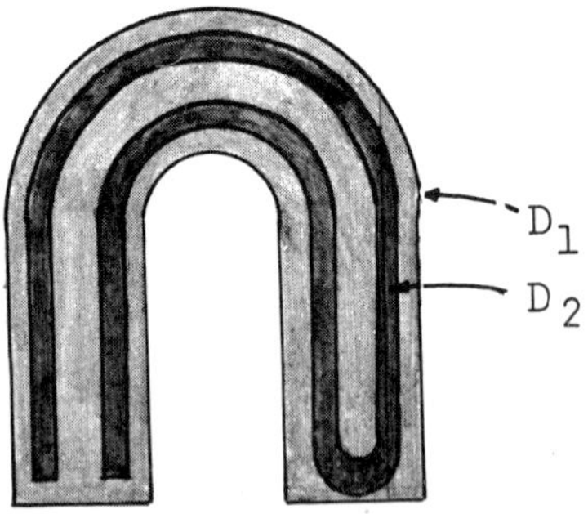

$\cap\{D_n : n = 1, 2, \ldots\}$

A continuum X is called _indecomposable_ iff whenever $X = X_1 \cup X_2$, X_1, X_2 continua, we have X_1 or $X_2 = X$. Further, X is _absolutely_ or _hereditarily indecomposable_ iff every subcontinuum is indecomposable.

Almost every continuum is absolutely indecomposable and snake-like. In fact, almost all of them are homeomorphic. (The last part is hard, see Bing [21].)

Actually, it is not hard to see that the snake-like continua are residual. Let E_ϵ denote the ϵ-snakes. Then E_ϵ is clearly open. Also, E_ϵ is dense. In fact if K is a continuum we can use an $\delta/2$ grid to find a simple arc within δ of K. The idea is as follows:

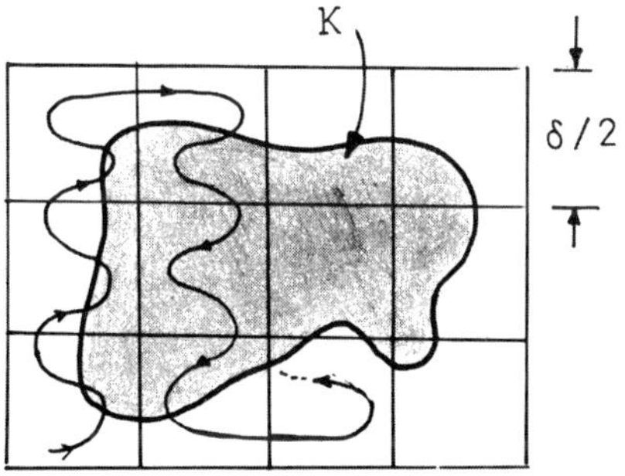

Then $E = \bigcap\{E_\epsilon\}$ is a residual set of snake like continua. The indecomposable case proceeds analogously.

One final comment. Although "first category" is analogous to "measure zero", there is in general no real link between them. See problem 7.2.

PROBLEMS FOR §7

7.1 Pointwise convergence of linear functions.

In general, convergence of a net of continuous functions must be uniform to ensure that the limit is continuous. Prove that in the case of a sequence of continuous linear functions on a Banach space, pointwise convergence already implies continuity of the limit function. Give an example to show that the convergence need not be uniform, nor in the norm.

7.2 A large first category set.

Show that there exists a first category set on $[0,1]$ of Lebesgue measure 1.

7.3 The dimensions of Banach spaces.

(i) Does there exist a Banach space whose linear dimension is countably infinite?

(ii) Let V be the space of polynomials $p: \mathbb{R} \to \mathbb{R}$. Is V Banachable?

7.4 Locally uniform convergence.

Let f_n be a sequence of real-valued continuous functions on $[0,1]$ converging pointwise to a map f. Show that there exists a residual set X in $[0,1]$ at each point of which, the sequence (f_n) converges uniformly (in a sense to be made precise).

7.5 Separate and joint continuity.

Let E be a topological space, F, G two metric spaces, d, d' the metrics in F, G respectively. Let f be a mapping of $E \times F$ into G such that, for each $x_0 \in E$, $y \mapsto f(x_0, y)$ is continuous in F, and such that, for each $y_0 \in F$, $x \mapsto f(x, y_0)$ is continuous in E.

(i) For each $\varepsilon > 0$, each $b \in F$ and each $x \in E$, let $g(x; b, \varepsilon)$ be the least upper bound of numbers $\alpha > 0$ such that: $(d(b,y) < \alpha) \Rightarrow (d'(f(x,b), f(x,y)) \leq \varepsilon.)$

Show that $x \mapsto g(x; b, \varepsilon)$ is an upper semi-continuous function.

(ii) If E is a Baire space, deduce from (i) that for each $b \in F$, there exists in E a set S_b whose complement is of 1st category, such that for each $a \in S_b$, f is continuous at the point (a,b).

(iii) Find out single conditions on E, F, G such that the set S_b may be chosen independent of b.

7.6 Products of normal spaces.

Let E be the set of real numbers with topology generated by the base of all sets of the form $[a,b[$.

(i) Show that E is regular.

(ii) Show that E is normal.

(iii) Show that $E \times E$ is not normal.

§8 CLASSIFICATION OF SETS AND FUNCTIONS

This section deals with some basic properties of Borel sets and functions. We also study the related concept of $\mathcal{K}$-analytic (or Souslin) sets. One of the main results, 8.7 gives the connection between strongly α-favorable and topologically complete spaces, which will be of use later in the book.

A good reference for much of this material is Kuratowski [1] Topology, (I), §§30-32.

8.1 DEFINITION. Let E be a set and $\mathcal{A} \subset \mathcal{P}(E)$. Recall that $\mathcal{A}$ is called a σ-field iff

(i) $X \in \mathcal{A}$ implies $E \setminus X \in \mathcal{A}$,

and (ii) $(X_n \in \mathcal{A}, \quad n = 1,2,\ldots) \Rightarrow (\cup\{X_n\} \in \mathcal{A})$.

For $\mathcal{B} \subset \mathcal{P}(X)$, the σ-field generated by $\mathcal{B}$ is $\cap\{\mathcal{A} \subset \mathcal{P}(X): \mathcal{A} \supset \mathcal{B}, \mathcal{A} \text{ is a } \sigma\text{-field}\}$. (Clearly this is the least σ-field containing $\mathcal{B}$).

Let E, F be sets with σ-fields $\mathcal{A} \subset \mathcal{P}(E)$ and $\mathcal{B} \subset \mathcal{P}(F)$ respectively. A mapping $f: E \to F$ is called measurable with respect to $\mathcal{A}$, $\mathcal{B}$ iff $(Y \in \mathcal{B} \Rightarrow f^{-1}(Y) \in \mathcal{A})$. (One can easily define initial σ-fields with respect to mappings $f_i: E \to E_i$;

this gives in particular _product_ σ-fields.)

Let E be a topological space with topology $\mathfrak{J}$. The σ-field generated by $\mathfrak{J}$ (or equivalently by the closed sets) is called the σ-field of _Borel sets_.

Suppose that $\mathfrak{F}$ denotes the family of closed sets on E, a topological space. We define inductively

$$\mathfrak{F},\ \mathfrak{F}_{\sigma},\ \mathfrak{F}_{\sigma\delta},\ \mathfrak{F}_{\sigma\delta\sigma},\ \ldots$$

where $\mathfrak{F}_{\sigma}$ denotes all countable unions of elements of $\mathfrak{F}$, $\mathfrak{F}_{\sigma\delta}$ denotes all countable intersections of members of $\mathfrak{F}_{\sigma}$, and so on.

Clearly $\mathfrak{F} \subset \mathfrak{F}_{\sigma} \subset \mathfrak{F}_{\sigma\delta} \subset \ldots,$ but the union of these need not be closed under countable unions and intersections. Below we shall obtain the least such family containing $\mathfrak{F}$, by continuing this process transfinitely.

Recall that σ-fields are closed under countable unions and intersections and that the Borel sets are the elements of the smallest family closed under countable unions and intersections which contains both the open and closed sets.

Let A be a non denumerable well ordered set such that for each $\alpha \in A$, $\{x \in A: x \leq \alpha\}$ is countable. We call elements of A, <u>ordinals of the second class</u> or more precisely, for each $\alpha \in A$, $\{x \in A: x \leq \alpha\}$ is an ordinal of the second class. For $\alpha \in A$ define the <u>successor</u> of α by $\alpha + 1 = \inf\{\beta \in A: \beta > \alpha\}$ and let $\alpha + n$ for n a non-negative integer be defined inductively. Also set $\beta_0 = \inf\{\beta \in A: \beta + n = \alpha$ for some non-negative integer $n\}$. We see that each α can be written uniquely as $\alpha = \alpha_0 + n$ where α_0 has no <u>predecessor</u> (an element β such that $\beta + 1 = \alpha_0$) and n an integer. We call α <u>even</u> or <u>odd</u> according as n is even or odd. (See §1).

By transfinite induction we can define families $\mathfrak{F}_\alpha$, where $\alpha = \alpha_0 + n \in A$ by

(i) if $\alpha = \alpha_0$ then $\mathfrak{F}_\alpha = \{\bigcap_{n=1}^{\infty} F_n: F_n \in \mathfrak{F}_\beta, \beta < \alpha\}$

(ii) $\mathfrak{F}_\alpha = (\mathfrak{F}_{\alpha_0})_{\sigma\delta\ldots\sigma}$ if α is odd

(iii) $\mathfrak{F}_\alpha = (\mathfrak{F}_{\alpha_0})_{\sigma\delta\ldots\delta}$ if α is even, where

$\sigma\delta\sigma\ldots$ is iterated n times.

Clearly $\alpha \leq \beta$ implies $\mathfrak{F}_\alpha \subset \mathfrak{F}_\beta$.

8.2 PROPOSITION. *Let* E *be a topological space and* $\mathfrak{F}$ *the class of closed sets. Consider (the net)*

$$\mathfrak{F} = \mathfrak{F}_0 \subset \mathfrak{F}_\sigma \subset \ldots \subset \mathfrak{F}_\alpha \subset \ldots$$

for α *any ordinal of the second class, as defined above. Then* $\bigcup\{\mathfrak{F}_\alpha\}$ *is the smallest family of closed sets which is closed under countable intersections and unions and which contains* $\mathfrak{F}$.

Proof. Let $X_1, \ldots, X_n, \ldots \in \bigcup\{\mathfrak{F}_\alpha\}$. Then we must show that $\bigcup\{X_n\} \in \bigcup\{\mathfrak{F}_\alpha\}$ and $\bigcap\{X_n\} \in \bigcup\{\mathfrak{F}_\alpha\}$. But there is an α such that $X_1, \ldots, X_n, \ldots \in \mathfrak{F}_\alpha$ as every countable subset of A has an upper bound. But if α is odd, $\bigcup\{X_n\} \in \mathfrak{F}_\alpha$ and $\bigcap\{X_n\} \in \mathfrak{F}_{\alpha+1}$ and vice-versa if α is even. □

Similar statements hold for $\mathcal{G}$, the class of open sets, and $\mathcal{U}$, the class of open or closed sets, where

$$\mathcal{G} = \mathcal{G}_0 \subset \mathcal{G}_\delta \subset \mathcal{G}_{\delta\sigma} \subset \ldots \subset \mathcal{G}_\alpha \subset \ldots$$

(for α even $\mathfrak{F}_\alpha$ is closed under countable intersections (multiplicative) and $\mathcal{G}_\alpha$ is closed under countable unions (additive) and vice-versa if α is odd).

8.3 PROPOSITION. *Suppose* E *has the property that* $\mathscr{G}$, *the class of open sets, lies in* $\mathfrak{F}_\sigma$ (*this holds when* E *is a metric space for example*). *Then for all* $\alpha \in A$,

$$\mathfrak{F}_\alpha \subset \mathscr{G}_{\alpha+1} \subset \mathfrak{F}_{\alpha+2} .$$

Further, $\cup\{\mathfrak{F}_\alpha\} = \cup\{\mathscr{G}_\alpha\}$ *is exactly the Borel sets on* E .

Proof. Note that $\mathscr{G} \subset \mathscr{F}_\sigma$ is equivalent to $\mathfrak{F} \subset \mathscr{G}_\delta$, so that our assertion is clear from the fact that $A \subset B$ implies $A_\alpha \subset B_\alpha$. The last statement follows from our remark that $\cup\{\mathfrak{K}_\alpha\}$ coincide with the Borel sets ($\mathfrak{K} = \mathfrak{F} \cup \mathscr{G}$).

To see that $\mathscr{G} \subset \mathfrak{F}_\sigma$ in a metric space, let U be open and $F_n = \{x \in E\colon d(x,E\setminus U) \geq 1/n\}$. Then $U = \cup\{F_n\}$. □

8.4 DEFINITION. Let E be a topological space with $\mathscr{G} \subset \mathfrak{F}_\sigma$ and X a Borel set. We say X is of *multiplicative class* α iff $X \in \mathfrak{F}_\alpha$ if α is even or $X \in \mathcal{G}_\alpha$ if α is odd (these are exactly the families closed under countable intersections).

Similarly X is of <u>additive class</u> α iff $X \in \mathcal{G}_\alpha$ if α is even and $X \in \mathfrak{F}_\alpha$ if α is odd.

Let E, F be topological spaces and $f: E \to F$ a mapping. We call f a <u>Borel function</u> iff for every $U \subset F$, U open, $f^{-1}(U)$ is a Borel set in E. Suppose on E, $\mathcal{G} \subset \mathfrak{F}_\sigma$ and $g: E \to F$ is a mapping. We say g is of (Baire) <u>class</u> α iff for each $X \subset F$ closed, $g^{-1}(X)$ is of multiplicative class α in E (or for U open, $g^{-1}(U)$ is of additive class α).

Although every Borel set is of some class α, a Borel function need not be. We do however have the following: <u>A Borel function</u> $f: E \to F$ <u>is of some class</u> α <u>whenever</u> F <u>has a denumerable base</u>. (<u>Proof</u>. Let $\mathfrak{B}$ denote the countable open base. Let α_V be the additive order associated with $f^{-1}(V)$, $V \in \mathfrak{B}$. Since $\mathfrak{B}$ is countable, there exists an ordinal β such that $\beta \geq \alpha_V$ for all $V \in \mathfrak{B}$. It follows that f is of class β. □)

Notice that g is of class 0 iff g is continuous and g is of class one iff $g^{-1}(X) \in \mathcal{G}_\delta$

for all closed sets (see 7.5).

8.5 THEOREM. Let E be a topological space and F a metric space with $f_n: E \to F$ mappings.

(i) if each f_n is of class α and $f_n \to f$ uniformly then f is of class α

(ii) if each f_n is of class α and $f_n \to f$ pointwise then f is of class $\alpha + 1$.

Actually, in (ii), the key property for F is the following:

8.6 LEMMA. Let F be a topological space such that for each $X \subset F$ closed, there exists open neighborhoods U_n of X such that $X = \bigcap\{c\ell(U_n)\}$. Then if $y_n \to b$, we have that: $(b \in X) \Longleftrightarrow$ (for all n there is a p so that $y_{n+p} \in U_n$).

We leave this easy lemma as an exercise.

Proof of 8.5. Let $X \subset F$ be closed and U_n open neighborhoods of X with $\bigcap\{c\ell(U_n)\} = X$. Then

$$f^{-1}(X) = \bigcap_{n=1}^{\infty} \{\bigcup_{p=1}^{\infty} f_{n+p}^{-1}(U_n)\}$$

by the lemma. But if each $f_{n+p}^{-1}(U_n)$ is of additive class α, $f^{-1}(X)$ is of multiplicative class $\alpha + 1$. This proves (ii).

For (i), choose a sequence k_n such that

$$d(f(x), f_{k_n+p}(x)) < 1/n \quad \text{for all} \quad x \in E,\ p \geq 0 .$$

Then for X closed,

$$f^{-1}(X) = \bigcap_{n=1}^{\infty} \bigcap_{p=1}^{\infty} f_{k_n+p}^{-1}(c\ell(U_n))$$

where $c\ell(U_n) = \{x \in F: d(x,X) \leq 1/n\}$. Hence if each f_n is of class α so is f. □

See Kuratowski, [1] Topology for more results along these lines.

The next theorem shows the basic role played by $\mathcal{G}_\delta$ sets.

8.7 THEOREM. Let (E,d) be a complete metric space and $X \subset E$. Then the following are equivalent:

(i) X is strongly α-favorable

(ii) X is a $\mathcal{G}_\delta$ in E

(iii) X is homeomorphic to a complete metric space. (That is, X is topologically complete).

Proof. (i) $\Rightarrow$ (ii). Let $B(x,\epsilon) = \{y \in E : d(x,y) < \epsilon\}$ and $S(x,\epsilon) = B(x,\epsilon) \cap X$. Let f be a winning tactic for X. For $n = 1,2,\ldots,$ let

$$U_n = \cup\{V' \subset E;\ V' \text{ is open, } V = V' \cap X$$

$$= f(S(x,1/n),x), \text{ and } V' \subset B(x,1/n) \text{ for some } x \in X\}$$

We claim that $X = \cap\{U_n\}$ which will prove the assertion (ii). That $X \subset \cap\{U_n\}$ is obvious (this would not follow if E were merely α-favorable.) Conversely, let $x \in \cap\{U_n\}$, and consider the following "game" in X. Let $n_1 = 1$ and $x \in V_1'$ where $V_1 = V_1' \cap X = f(S(y_1,1/n_1),y_1)$ for some $y_1 \in X$. Let $\epsilon = d(x,E\setminus V') > 0$ and choose n_2 such that $n_2 > n_1$ and $1/n_2 < \epsilon/2$. Since $x \in U_{n_2}$, $x \in V_2'$ where $V_2 = V_2' \cap X = f(S(y_2,1/n_2,y_2))$ for some $y_2 \in X$. Notice that $S(y_2,1/n_2) \subset V_1$ (since $y \in S(y_2,1/n_2)$ implies that $d(x,y) \leq d(x,y_2) + d(y_2,y) < 1/n_2 + 1/n_2 < \epsilon$ by the facts that $x \in V_2' \subset B(y_2,1/n_2)$ and $S(y_2,1/n_2) \subset B(y_2,1/n_2)$, $d(x,y) < \epsilon$ implies that $y \in V_1'$, and hence $y \in V_1$). Inductively we define (in the same way) a game

$S(y_1,1/n_1) \supset f(S(y_1,1/n_1),y_1) \supset S(y_2,1/n_2)$ $\supset f(S(y_2,1/n_2),y_2) \ldots$ and since f is a winning tactic, $\cap\{V_n\} \neq \emptyset$. But since $n_1 < n_2 < \ldots$ we have $\{x\} = \cap\{V_n'\} \supset \cap\{V_n\}$, and hence $\{x\} = \cap(V_n)$ and in particular, $x \in X$.

(ii) $\Rightarrow$ (iii). Suppose that $X = \cap\{U_n\}$ where the $U_n \subset E$ are open, $n = 1,2,\ldots$. For each n, define, for $x \in U_n$,

$$f_n(x) = 1/d(x,E\setminus U_n) .$$

Let

$$d_n(x,y) = d(x,y) + |f_n(x) - f_n(y)| \text{ for } x, y \in U_n.$$

Since $d \leq d_n$, the d_n topology (on U_n) is finer than the d topology. But since, for each $y \in U_n$, $g_y(x) = d_n(x,y)$ is continuous in the d topology, they generate the same topology. Also, the space (U_n,d_n) is complete since a d_n-Cauchy sequence is also d-Cauchy and hence converges in E. But the limit must lie in U_n, for if $x_n \to x \in bd(U_n)$, then $f_n(x_n) \to \infty$.

Now by 5.25, $\Pi\{U_n\}$ is a complete metric space. Define $\Phi: X \to \Pi\{U_n\}$ by $(\Phi(x))_n = x$.

Clearly Φ is one to one and is continuous, since for any projection π_n, $\pi_n \circ \Phi$ is the identity map of X onto X. Also, Φ^{-1} is continuous, for if V is open in X, $V = X \cap U$ for some U open in E and $\Phi(V) = \pi_n^{-1}(U) \cap \Phi(X)$ for any n. Thus Φ is a homeomorphism onto its image. It suffices to show that the image is closed. However, since $X = \bigcap\{U_n\}$, $\Phi(X)$ is the diagonal of $\Pi\{U_n\}$, so is closed.

(iii) $\Rightarrow$ (i). This follows at once from 7.12 and the fact that the property "strongly α-favorable" is a topological property (invariant under homeomorphisms). □

<u>Remarks</u>. The implication (i) $\Rightarrow$ (ii) does not use the hypothesis that d is complete. For (E,d) one can use the completion of X, for example.

In the proof of (i) $\Rightarrow$ (ii) we were given a strategy which depended only on the previous β move. However, one can show, by modifying the game, that we could have used a strategy which depended on the previous moves. Thus if we allowed α's strategy in the definition of strongly α-favorable to depend on the previous moves, the theorem would still be true.

This remark is the basis of the following important result of F. Hausdorff [1]. See also Frolik [1].

8.8 <u>COROLLARY</u>. <u>If</u> E <u>is a topologically complete space</u>, F <u>is a metrizable space, and</u> $f: E \to F$ <u>is an open continuous surjection, then</u> F <u>is topologically complete</u>.

<u>Proof</u>. (Assuming our above remarks) We may assume that E is a complete metric space. We shall show that F has an α-winning tactic based on the previous α and β choices. Suppose the nth choice of α in F is $f(\omega_n)$ where ω_n is open in E, and the succeeding β choice is (Ω_{n+1}, y_n) for Ω_{n+1} open in F. Choose $c\ell(\omega_{n+1}) \subset \omega_n \cap f^{-1}(\Omega_{n+1})$ and ω_{n+1} containing some point of $f^{-1}(y_n)$, and such that diameter $(\omega_{n+1}) \leq \inf(1, \text{diameter } \omega_n)/$ This gives a winning tactic on F. □

Next we turn to the basic properties of $\mathcal{K}$-analytic sets and Souslin sets. References are G. Choquet [2] and D.W. Bressler and M. Sion [1] and

Sierpinski, <u>General Topology</u>.

8.9 <u>DEFINITION</u>. Let E be a Hausdorff space and $\mathcal{K}$ the family of compact sets in E. The smallest family containing $\mathcal{K}$ and closed under countable unions and intersections is called the family of <u>$\mathcal{K}$-Borel sets in</u> E. (Thus the $\mathcal{K}$-Borel sets are $\cup\{\mathcal{K}_\alpha : \alpha$ is an ordinal of the second class$\}$ as described in 8.2). Alternatively, the $\mathcal{K}$-Borel sets form the monotone class generated by $\mathcal{K}$. (See §3).

A set $A \subset E$ is called a $\mathcal{K}$-<u>analytic</u> set iff there exists a compact space F, a $\mathcal{K}_{\sigma\delta}$ set $B \subset F$ and a map $f: B \to E$ which is continuous and $f(B) = A$.

The $\mathcal{K}$-analytic sets are designed to be stable under continuous mappings and also to be a large class of sets. If we had replaced $\mathcal{K}_{\sigma\delta}$ by $\mathcal{K}$ or $\mathcal{K}_\sigma$, it is obvious that we would have obtained as $\mathcal{K}$-analytic sets merely $\mathcal{K}$ and $\mathcal{K}_\sigma$ in E, but with $\mathcal{K}_{\sigma\delta}$ we get "new" sets, as we shall see below.

Recall that for a family of spaces $\{X_i\}$, the <u>topological sum</u> $\Sigma\{X_i\}$, is the disjoint union with the final topology relative to the inclusion maps $f_i: X_i \to \Sigma\{X_i\}$.

8.10 PROPOSITION. (i) *Suppose* $f: E \to F$ *is continuous and* $A \subset E$ *is a* $\mathcal{K}$-*analytic set. Then* $f(A) \subset F$ *is a* $\mathcal{K}$-*analytic set*

(ii) *if* $X_1, X_2, \ldots$ *are* $\mathcal{K}$-*analytic sets then so are* $\bigcup\{X_n\}$ *and* $\bigcap\{X_n\}$.

Proof. (i) is obvious. For (ii), let $f_n: B_n \subset F_n \to E$ where B_n is a $\mathcal{K}_{\sigma\delta}$ in the compact space F_n and $f_n(B_n) = X_n$. Consider $F_0 = \Sigma F_n$ and let F be a compact space containing F_0 (for example the one point compactification (§2)). Then let $B = \Sigma\{B_n\}$ and define $f: B \to E$ by $f(x) = f_n(x)$ if $x \in B_n$. Note that $B = \bigcap(B_n \cup [F \setminus F_n])$, $B \in \mathcal{K}_{\sigma\delta}$ and $f(B) = \bigcup X_n$. This proves that $\bigcup X_n$ is $\mathcal{K}$-analytic. For the intersection, take $F = \Pi F_n$ and $B' = \Pi B_n$; then $B' \in \mathcal{K}_{\sigma\delta}$ in F . Let

$$B'_{m,n} = \{x \in B': f_m \circ \pi_m(x) = f_n \circ \pi_n(x)\}$$

where $\pi_n: F \to F_n$ is the projection. Then as E is Hausdorff, $B'_{m,n}$ is closed in B , and so is a $\mathcal{K}_{\sigma\delta}$ set in F . Let $B = \bigcap\{B'_{m,n}: m,n = 1,2,\ldots\}$ We can then define $f: B \to E$ as the common value

$f_n \pi_n(x)$. Then f is continuous and $f(B) \subset \cap X_n$. Conversely, if $x \in \cap X_n$, $x = f_1(x_1) = \ldots = f_n(x_n) = \ldots$ so that $f(B) \supset \cap X_n$. □

Property (i) for $\mathcal{K}$-analytic sets is one of their basic advantages over measurable sets.

8.11 COROLLARY. *Let* E *be a Hausdorff space. Then every* $\mathcal{K}$-*Borel set is* $\mathcal{K}$-*analytic*.

Proof. The $\mathcal{K}$-analytic sets contain the compact sets and are closed under countable intersections and unions. Hence they contain the $\mathcal{K}$-Borel sets. □

Except for trivial cases, the $\mathcal{K}$-analytic sets properly contain the $\mathcal{K}$-Borel sets.

★The Classical Theory of Analytic Sets

Let us now briefly compare this with the "classical" theory of analytic sets. We shall need the following fact: if $I \subset [0,1]$ denotes the irrationals and $\mathbb{N} = \{1,2,3,\ldots\}$ with the discrete topology, then I and $\mathbb{N}^{\mathbb{N}}$ are homeomorphic. The homeomorphism is $\Psi(n_1,n_2,\ldots) = 1/(n_1 + 1/(n_2 + 1/(n_3 + \ldots$; this is a result from the theory of continued fractions).

8.12 DEFINITION. Let E be a Hausdorff topological space and $A \subset E$. We say A is an *analytic set* (sometimes called a *Souslin set*) iff there exists a continuous map $f\colon \mathbb{N}^{\mathbb{N}} \to E$ with $f(\mathbb{N}^{\mathbb{N}}) = A$.

8.13 PROPOSITION. *Every analytic set in E is $\mathcal{K}$-analytic.*

Proof. First, I is a $\mathcal{K}_{\sigma\delta}$ in $[0,1]$. (In fact if $r_1, r_2, \ldots$ are the rationals, $[0,1]\setminus\{r_n\}$ is a $\mathcal{K}_\sigma$ and $I = \bigcap\{[0,1]\setminus\{r_n\}\}$.) Hence if $A = f(\mathbb{N}^{\mathbb{N}})$, $A = f \circ \Psi^{-1}(I)$, so A is a $\mathcal{K}$-analytic set. □

Basic properties of analytic sets are:

8.14 PROPOSITION. (i) *Suppose $f\colon E \to F$ is a continuous mapping and $A \subset E$ is analytic. Then $f(A)$ is analytic.*

(ii) *If $X_1, X_2, \ldots$ are analytic sets in E then $\bigcup\{X_n\}$ and $\bigcap\{X_n\}$ are analytic sets.*

Proof. (i) is clear. For (ii), let $f_n\colon \mathbb{N}^{\mathbb{N}} \to E$ have $f_n(\mathbb{N}^{\mathbb{N}}) = X_n$. Define $f\colon \mathbb{N}^{\mathbb{N}} \to E$ by

$f(a_1,a_2,\ldots) = f_{a_1}(a_2,a_3,\ldots)$; this is clearly continuous. Obviously $f(\mathbb{N}^{\mathbb{N}}) = \bigcup\{X_n\}$, so that $\bigcup\{X_n\}$ is an analytic set.

For the second part we first prove:

8.15 LEMMA. *Every closed subset of* $\mathbb{N}^{\mathbb{N}}$ *is analytic*.

Proof. Notice that $\mathbb{N}^{\mathbb{N}}$ is a complete separable metric space by 5.25. (In fact, a metric is $d(a,b) = \sum_{i=1}^{\infty} 2^{-i}|a_i - b_i|/\{1 + |a_i - b_i|\}$; see Kelley, *General Topology*, p. 120.)

Inductively, for each closed set F in $\mathbb{N}^{\mathbb{N}}$ write

$$F = \bigcup\{F_{i_1} : i_1 = 1,2,\ldots\}$$

$$F_{i_1\ldots i_k} = \bigcup\{F_{i_1\ldots i_{k+1}} : i_{k+1} = 1,2,\ldots\}$$

where each $F_{i_1\ldots i_k}$ is closed of diameter $< 1/2^k$.

Given $a = (a_1,\ldots) \in \mathbb{N}^{\mathbb{N}}$ define

$$\varphi(a) = \bigcap\{F_{a_1\ldots a_i} : i = 1,2,\ldots\} \in F$$

which is exactly one point. Then φ is clearly a

map onto F. For continuity of φ notice that if

$$2^{-k} \le d(a,b) < 2^{-k+1}$$

then $a_i = b_i$ for $i \ge k$, so that

$$d(\varphi(a), \varphi(b)) \le \text{diameter } (F_{a_k \ldots a_k}) \le 2^{-k}.$$

Thus φ is continuous (in fact Lipschitz). □

To complete the proof of 8.14, suppose as before that $f_n: \mathbb{N}^{\mathbb{N}} \to E$ has $f_n(\mathbb{N}^{\mathbb{N}}) = X_n$. Form $(\mathbb{N}^{\mathbb{N}})^{\mathbb{N}} \approx \mathbb{N}^{\mathbb{N}}$ (since $\mathbb{N} \times \mathbb{N}$ is countable). Let $B_{n,m} = \{x \in (\mathbb{N}^{\mathbb{N}})^{\mathbb{N}} : f_n \quad \pi_n(x) = f_m \quad \pi_m(x)\}$ and $B = \bigcap\{B_{n,m}\}$ as in 8.10. In view of the lemma we have the result. □

8.16 PROPOSITION. *If E is a σ-compact metric space (and therefore separable) then every Borel set is analytic.*

Proof. By 8.14 we may assume E is compact. From 6.3, E may be embedded as a closed set in $[0,1]^{\mathbb{N}}$. Now $[0,1]$ is analytic since, if $r_1, r_2, \ldots$ are the rationals, $[0,1] = \bigcup\{\{r_k\} \cup I\}$. Therefore we have a map

$$h: I \approx I^{\mathbb{N}} \to [0,1]^{\mathbb{N}} .$$

If $A \subset [0,1]^{\mathbb{N}}$ is Borel, then $h^{-1}(A)$ is a Borel set in I. Therefore it will be sufficient to prove the following:

8.17 <u>LEMMA</u>. <u>Any Borel set in</u> I <u>is analytic.</u>

<u>Proof</u>. By 8.15 every closed set in I is analytic. Since every open set is an $\mathfrak{F}_\sigma$, the open sets are also analytic. Therefore by 8.14 the analytic sets contain the Borel sets. □

Since $\mathbb{N} \approx \mathbb{N} \times \mathbb{N}$, we also see that the countable product of analytic sets is also analytic.

The following result shows that in most interesting cases the notions of $\mathcal{K}$-analytic set and analytic sets coincide:

8.18 <u>THEOREM</u>. <u>In a metric space the</u> $\mathcal{K}$-<u>analytic sets coincide with the analytic sets</u>.

This is not needed later so we omit the proof. It may be found in G. Choquet [3].

PROBLEMS FOR §8

8.1 Cardinality of σ-fields.

Show that on any set, every infinite σ-field is uncountable.

8.2 $\mathcal{K}$-Borel and Borel Sets.

(i) If E is σ-compact, the σ-field generated by the compact sets coincides with the Borel sets.

(ii) If E is locally compact Hausdorff with countable base, the $\mathcal{K}$-Borel and Borel sets coincide.

8.3 Analytic subsets of $\mathcal{C}([0,1],\mathbf{R})$.

Consider the Banach space $\mathcal{C} = \mathcal{C}([0,1],\mathbf{R})$.

(i) Show that every compact subset of $\mathcal{C}$ is nowhere dense.

(ii) Show that the set $\mathcal{L}$ of Lipschitz functions is a $\mathcal{K}_\sigma$ in $\mathcal{C}$.

(iii) Show that the set $\mathcal{C}^1 \subset \mathcal{C}$ of functions with a continuous derivative is an analytic subset of $\mathcal{C}$. Is it a Borel set?

(iv) What can you say about $\mathcal{C}^\infty \subset \mathcal{C}$, the infinitely differentiable functions?

8.4 Extension of Homeomorphisms.

(i) Let E and F be metric spaces with F complete. Suppose $f: X \subset E \to F$ is continuous. Show that f may be extended as a continuous map to a $\mathscr{G}_\delta$ set containing X .

(ii) Let E and F be complete metric spaces and $f: X \subset E \to Y \subset F$ a homeomorphism of X onto Y . Then f may be extended to a homeomorphism between $\mathscr{G}_\delta$ sets containing X and Y .

(iii) Use this result to give an alternative proof of (iii) $\Rightarrow$ (ii) in 8.7.

CHAPTER 3

RADON MEASURES

In this chapter we study Radon measures from the Bourbaki point of view. This point of view is important, for it allows us to apply powerful theorems on topological vector spaces (whose proofs are given later) to deduce useful results about Radon measures.

The first section of this chapter is not central to our presentation but is included for completeness. Capacities are important in potential theory as well as for theorems about Radon measures, and so are of independent interest. The main theorems on Radon measures are given in §11, and §12 and in that context, the usual operations on measures are discussed in the subsequent sections.

★ §9 CAPACITIES

The notion of capacity arose in potential theory and has important applications there. (See Meyer, [1] Probability and Potentials.) It also provides a non-trivial generalization of standard arguments of measure theory. (For example, the tools can be used for the Daniell Integral; see Meyer, p. 45). The reader not interested in these tools may omit this section with no loss of continuity, (except for one important application in section 27).

An important example of a capacity is the Newtonian capacity for potential theory. The potential of a point mass m at $x \in \mathbb{R}^3$ is defined by $\varphi(y) = m/\|x-y\|$. More generally, the potential of a measure μ is

$$\varphi_\mu(y) = \int (1/\|x-y\|)d\mu(x) .$$

Instead of mass, one can think of μ as being a "charge density", in which case φ has a physical interpretation as potential energy at the point y. In physics there is a useful notion of capacity of a set (that is, of a "conductor") which is, defined on compacts, to be: capacity $(K) = \sup\{\mu(K): \mu$ is supported by K and $\varphi_\mu \leq 1\}$ (this is the "charge" $\mu(K)$ per unit "potential" φ_μ) .

These physical notions have evolved into the vast subject of potential theory.

Another example of a capacity is a regular Borel measure as defined in §3. This will be shown below.

A basic extension theorem for capacities is given in 9.7 and is the analogue of the outer measure in measure theory. The second basic theorem (9.3) of this section deals with a generalization of the concept of inner regularity of measures.

The main reference for this section is G. Choquet [4].

9.1 DEFINITION. Let E be a Hausdorff space. A (regular) capacity on E is a mapping $f: \mathcal{P}(E) \to \mathbb{R} \cup \{\pm \infty\}$ such that

(i) f is monotone (increasing): $(X \subset Y) \Rightarrow (f(X) \leq f(Y))$,

(ii) if $X_n \uparrow X$ is an increasing sequence of sets, then $f(X_n) \uparrow f(X)$.

(iii) if $K_n \downarrow K$ is a decreasing sequence

of compact sets, then $f(K_n) \downarrow f(K)$.

For a measure μ , recall that a set X is <u>inner regular</u> iff $\mu(X) = \sup\{\mu(K): K \subset X$ and K is compact$\}$; see §3. For capacities we make an analogous definition:

9.2 <u>DEFINITION</u>. Let f be a capacity on E . A subset $X \subset E$ is called f-<u>capacitable</u> iff

$$f(X) = \sup\{f(K): K \subset X \text{ and } K \text{ is compact}\} .$$

We shall prove the following result which, at first sight, may seem rather implausible:

9.3 <u>THEOREM</u>. <u>Let</u> E <u>be a Hausdorff space and</u> f <u>a capacity on</u> E . <u>Then every</u> $\mathcal{K}$-<u>analytic subset which lies in some</u> $\mathcal{K}_\sigma$ <u>set is</u> f-<u>capacitable</u>. <u>In particular, every</u> $\mathcal{K}$-<u>Borel set is</u> f-<u>capacitable</u>.

The proof of this theorem will follow from a number of lemmas. First, we show why the last statement follows. For this, it suffices to show that every $\mathcal{K}$-Borel set lies in some $\mathcal{K}_\sigma$. In fact, if $\mathcal{E}$ is any collection of subsets of a set E and $\mathcal{A}$ is the monotone class generated by $\mathcal{E}$, then every $X \in \mathcal{A}$ is contained in a countable union of elements of $\mathcal{E}$. (<u>Proof</u>. Let $\mathcal{J}$ denote all members of $\mathcal{A}$ contained in a countable union of elements of $\mathcal{E}$. Clearly $\mathcal{J}$ is a monotone class containing $\mathcal{E}$, and so $\mathcal{J} = \mathcal{A}$).

9.4 <u>LEMMA</u>. <u>Let</u> f <u>be a capacity on a Hausdorff space</u> F. <u>Then any</u> $\mathcal{K}_{\sigma\delta}$ <u>in</u> F <u>is</u> f-<u>capacitable</u>.

<u>Proof</u>. Compacts are obviously f-capacitable. If $B = \bigcup\{K_n\}$ is a $\mathcal{K}_\sigma$ then we may assume K_n are increasing and $f(B) = \lim_{n\to\infty} f(K_n) \leq \sup\{f(K): K \subset B, K \text{ compact}\} \leq f(B)$, so B is f-capacitable. Now let $A = \bigcap\{A_n\}$, where $A_n = \bigcup\{K_{n,p}: p = 1,2,\ldots\}$ and K_{np} is compact and increases with p. Choose $\lambda < f(A)$. Since $A \subset A_1$, the sets $A \cap K_{1,p}$ increase to A so there is a p_1 with $f(A \cap K_{1,p_1}) > \lambda$. Let $a_1 = A \cap K_{1,p_1}$. Define inductively, $a_n = a_{n-1} \cap K_{n,p_n}$ such that $f(a_n) > \lambda$. Since $a_n \subset K_{1,p_1} \cap \ldots \cap K_{n,p_n}$, we have $f(K_{1,p_1} \cap \ldots \cap K_{n,p_n}) > \lambda$. Let $K = \bigcap\{K_{n,p_n}\}$ so that $\lambda \leq f(K) \leq f(A)$. Hence A is f-capacitable. □

9.5 LEMMA. *Let* E *be a Hausdorff space and* A *a* $\mathcal{K}$-*analytic set in* E *which lies in a* $\mathcal{K}_\sigma$ *set. Then there is a* $\mathcal{K}_{\sigma\delta}$ *set* B *in a* σ-*compact space* F *and a continuous mapping* $f: F \to E$ *such that* $f(B) = A$.

Proof. Suppose $g: B \subset F' \to E$ has $g(B) = A$, B a $\mathcal{K}_{\sigma\delta}$ in a compact space F' . Let γ denote the graph of g . Now $c\ell(\gamma)$ is σ-compact and the projection $f: F' \times E \to E$ has $f(\gamma) = A$. But $\gamma = c\ell(\gamma) \cap (B \times E)$ is a $\mathcal{K}_{\sigma\delta}$ in $c\ell(\gamma)$ so that $f: c\ell(\gamma) \to E$ is the required map with γ the $\mathcal{K}_{\sigma\delta}$, and $F = c\ell(\gamma)$. □

9.6 LEMMA. *Let* E *be a Hausdorff space,* $\varphi : F \to E$ *a continuous map and* f *a capacity on* E . *Then* $g: \mathcal{P}(F) \to \mathbb{R} \cup \{\pm\infty\}$, *defined by* $g(X) = f(\varphi(X))$ *is a capacity on* F . *If* X *is* g-*capacitable in* F , *then* $\varphi(X)$ *is* f-*capacitable in* E .

This follows easily from the definitions.

Proof of 9.3. By 9.5, if A is a $\mathcal{K}$-analytic set, A is the image of a $\mathcal{K}_{\sigma\delta}$, B , in a σ-compact space

F under a map $\varphi: F \to E$. By 9.4, B is $f \circ \varphi$-capacitable, and by 9.6, A is f-capacitable. □

An interesting particular case is that of an f which is initially defined only on a subset of $\mathcal{P}(E)$; for example on $\mathcal{K}$, the compact sets. In such a situation we have an extension problem; that is, to find sufficient conditions on f to guarantee an extension to all of $\mathcal{P}(E)$. The procedure is entirely analogous to regularizing a Borel measure in measure theory.

The main theorem is the following:

9.7 THEOREM. Let E be a Hausdorff space and $\mathcal{K}$ the class of compact sets. Suppose $f: \mathcal{K} \to \mathbb{R}^+ \cup \{\infty\} = \{x \in \mathbb{R} \cup \{\infty\} : x \geq 0\}$ satisfies the following conditions

(i) f is increasing,

(ii) (continuity on the right) for $K \in \mathcal{K}$, $\varepsilon > 0$ there is an open set $U \supset K$ such that if $K' \in \mathcal{K}$ and $K \subset K' \subset U$ we have $f(K') < f(K) + \varepsilon$,

(iii) (strong subadditivity) for $K_1, K_2 \in \mathcal{K}$,

$$f(K_1 \cup K_2) + f(K_1 \cap K_2) \leq f(K_1) + f(K_2).$$

Suppose f^* is defined on $\mathcal{P}(E)$ by

$$f^*(X) = \inf\{f_*(U): U \supset X , U \text{ open}\}$$

(the outer capacity), where

$$f_*(X) = \sup\{f(K): K \subset X , K \text{ compact}\}$$

(the inner capacity).

Then f^* is a (regular) capacity coinciding with f on $\mathcal{K}$. In particular, $f^*(A) = f_*(A)$ for every $\mathcal{K}$-analytic (or $\mathcal{K}$-Borel) set A, lying in a $\mathcal{K}_\sigma$ by 9.3.

Observe that f^* and f_* coincide on the open sets.

For the proof we prepare the following lemmas:

9.8 LEMMA. Let $f: \mathcal{K} \to \mathbb{R}^+ \cup\{\infty\}$ satisfy (i) and (ii) above. Then f_* and f^* are increasing and if $K_1 \supset K_2 \supset \ldots$ is a decreasing sequence of compact sets, $f^*(\bigcap\{K_n\}) = \lim_{n\to\infty} f^*(K_n)$. Furthermore, f_* and f^* coincide with f on $\mathcal{K}$.

Proof. Clearly f^* and f_* are increasing. We now show that if $\mathcal{K}$ is compact then

$f_*(K) = f(K) = f^*(K)$. The first equality is obvious. For the second equality, given $\varepsilon > 0$, choose an open set $U \supset K$ such that $K \subset K' \subset U$ implies $f(K') - f(K) < \varepsilon$. Thus, $f_*(U) \leq f(K) + \varepsilon$ and hence $f^*(K) \leq f(K) + \varepsilon$.

Finally we must show that for compacts $K_1 \supset K_2 \supset \ldots$, $K = \cap\{K_n\}$ implies that $f(K) = \lim_{n\to\infty} f(K_n)$. Choose $U \supset K$ so that $K \subset K' \subset U$ implies $f(K') - f(K) < \varepsilon$. But there exists N so that $K_n \subset U$ for $n \geq N$ by compactness. □

9.9 <u>LEMMA</u>. <u>Under the hypotheses of 9.7, if</u> U <u>and</u> V <u>are open then</u>

$$f^*(U \cap V) + f^*(U \cup V) \leq f^*(U) + f^*(V)$$

(<u>or equivalently, for</u> f_*) .

<u>Proof</u>. For K_1, K_2 compact in U_1, U_2 respectivel[y] we have by 9.7 (iii),

$$f(K_1 \cup K_2) + f(K_1 \cap K_2) \leq f_*(U) + f_*(V).$$

Given $\varepsilon > 0$ choose $K \subset U \cup V$, and $K' \subset U \cap V$ so that

$$f_*(U \cap V) - f(K') < \epsilon/2 ,$$

and
$$f_*(U \cup V) - f(K) < \epsilon/2 ,$$

where $K' \subset K$. Consider

$$H_1 = K \cap (E \setminus U)$$

and
$$H_2 = K \cap (E \setminus V) .$$

By normality of K there are disjoint open neighborhoods U_1, U_2 of H_1 and H_2 in K . Further, we may assume $U_1 \subset E \setminus K'$ and $U_2 \subset E \setminus K'$ as $K' \subset U_1 \cap U_2$. Let $K_1 = K \setminus U$, and $K_2 = K \setminus U_2$. Hence $K_1 \cup K_2 = K$ and $K_1 \cap K_2 \supset K'$. Therefore,

$$f(K) + f(K') \leq f(K_1 \cup K_2) + f(K_1 \cap K_2)$$
$$\leq f_*(U) + f_*(V)$$

since $K_1 \subset U$ and $K_2 \subset V$. Hence

$$f_*(U \cup V) + f_*(U \cap V) \leq f_*(U) + f_*(V) + \epsilon$$

which proves the result. □

This lemma is now generalized to the following:

9.10 LEMMA. _Under the hypotheses of 9.7, if_

$U_1,\ldots,U_n$, $V_1,\ldots,V_n$ <u>are open sets with</u> $U_i \subset U_i$ $i = 1,\ldots,n$, <u>then</u>

$$f^*(\bigcup\{U_i\}) + \Sigma\{f^*(V_i)\} \leq f^*(\bigcup\{V_i\}) + \Sigma\{f^*(U_i)\}.$$

<u>Proof</u>. We proceed by induction on n . The case $n = 1$ is trivial. For the case $n = 2$ we have, by 9.9,

$$f^*(U_1 \cup U_2) + f^*(V_1) \leq f^*(V_1 \cup U_2) + f^*(U_1)$$

using $U = U_1$, $V = U_2 \cup V_1$ and the fact that f^* is increasing. Also, from 9.9,

$$f^*(U_2 \cup V_1) + f^*(V_2) \leq f^*(V_1 \cup V_2) + f^*(U_2)$$

using $U = U_2$ and $V = V_1 \cup V_2$. Adding gives the result. The result for $n+1$ follows from the case $k \leq n$ by applying the case $n=2$ to the four sets

$\bigcup\{V_i\colon i = 1,\ldots,n\}, \bigcup\{U_i\colon 1 = 1,\ldots,n\}, V_{n+1}, U_{n+1}$. □

9.11 <u>LEMMA</u>. <u>Under the hypotheses of</u> 9.7, <u>if</u> $X_1,\ldots,X_n$, $Y_1,\ldots,Y_n$ <u>are subsets of</u> E <u>with</u> $Y_i \subset X_i$, <u>then</u>

$$f^*(\bigcup\{X_i\}) + \Sigma\{f^*(Y_i)\} \leq f^*(\bigcup\{Y_i\}) + \Sigma\{f^*(X_i)\}.$$

Proof. Suppose $\bigcup\{Y_i\} \subset V$, $X_i \subset U_i$; U_i, V open so that by 9.10 applied to $U_1,\ldots,U_n$, $V_i = V \cap U_i$,
$f^*(\bigcup\{X_i\}) + \Sigma\{f^*(Y_i)\} \leq f^*(\bigcup\{U_i\}) + \Sigma\{f^*(V_i)\}$
$\leq f^*(V) + \Sigma\{f^*(U_i)\}$ which gives the result by taking the infimum. □

9.12 LEMMA. Under the hypotheses of 9.7, suppose $U_1 \subset U_2 \subset \ldots$ is an increasing sequence of open sets. Then if $U_n \uparrow U$,

$$f^*(U) = \lim_{n\to\infty} f^*(U_n) .$$

Proof. Since f^* is increasing we have

$$\lim_{n\to\infty} f^*(U_n) \leq f^*(U) .$$

Let $\lambda < f^*(U)$ and choose a compact set $K \subset U$ with $f(K) > \lambda$. Now $\{K \setminus U_i\}$ is a family with void intersection and so by compactness, $K \subset U_n$ for some n. Hence $f^*(U_n) > \lambda$. □

Proof of 9.7. It remains to be shown that for an increasing family $X_n \subset E$, $X_n \uparrow X$ implies

$$f^*(X) = \lim_{n\to\infty} f^*(X_n) .$$

Clearly it suffices to show $f^*(X) \leq \lim_{n\to\infty} f^*(X_n)$. For each n choose $U_n \supset X_n$, U_n open with $|f^*(X_n) - f^*(U_n)| < \epsilon/2^n$. Then by 9.11,

$$f^*(\bigcup_{i=1}^{n} U_i) + \sum_{i=1}^{n} f^*(X_n) \leq f^*(\bigcup_{i=1}^{n} X_n) + \sum_{i=1}^{n} f^*(U_i)$$

or

$$f^*(\bigcup_{i=1}^{n} U_i) - f^*(\bigcup_{i=1}^{n} X_n) \leq \epsilon .$$

Letting $n \to \infty$ and using 9.12 gives

$$f^*(\bigcup_{i=1}^{\infty} U_i) \leq \lim_{n\to\infty} f^*(X_n) + \epsilon .$$

Hence

$$f^*(X) \leq \lim_{n\to\infty} f^*(X_n) + \epsilon . \square$$

The Newtonian capacity described at the beginning of this section satisfies the hypotheses of 9.7 and so may be extended to a (regular) capacity (this is not completely obvious).

If μ is regular Borel measure on a locally

compact Hausdorff space, then it is obvious that μ satisfies the conditions of 9.7 ((ii) by regularity and (iii) by additivity), so that μ^*, its outer measure is a capacity and all the $\mathcal{K}$-analytic sets are regular, that is, $\mu^*(A) = \mu_*(A)$, and in particular are measurable. Notice that the conditions of 9.1 then follow for μ^* even though the sets may not be measurable!

Other capacities which motivated the general theory are described in the following without proof.

9.13 EXAMPLE. Let E be a compact metric space and $\mathcal{K}$ the space of non-empty compact sets in E with the Hausdorff metric. For $K \in \mathcal{K}$ let $A(K) = \{F \in \mathcal{K} : F \subset K\}$ which is a closed subset of $\mathcal{K}$. Let μ be a positive Radon measure on $\mathcal{K}$, and define $f: \mathcal{K} \to R^+ \cup \{\infty\}$ by $f(K) = \mu(A(K))$.

Then f satisfies the conditions of 9.7. Notice however that although μ is additive, f is not.

More specifically if $E = [0,1] \times [0,1]$ and we consider "Brownian motion" on E, the probability that a trajectory hits $K \subset E$ before the boundary is a capacity of this type.

Much of what we have done for capacities can be generalized in an abstract setting and in fact this generalization is useful. The basic facts are sketched below:

9.14 DEFINITION. Let E be a set and $\mathcal{K} \subset \mathcal{P}(E)$, such that $\mathcal{K}$ is stable under finite unions and denumerable intersections ($X_1, \ldots, X_n, \ldots \in \mathcal{K}$ implies $\bigcup_{n=1}^{N} X_n \in \mathcal{K}$ and $\bigcap_{n=1}^{\infty} X_n \in \mathcal{K}$). An $\mathcal{K}$-capacity on E is a mapping $f: \mathcal{P}(E) \to \mathbb{R} \cup \{\pm\infty\}$ such that

(i) f is increasing

(ii) for any increasing sequence X_n, $X_n \in \mathcal{P}(E)$, $X_n \uparrow X$,

$$f(X) = \lim_{n\to\infty} f(X_n)$$

(iii) if $K_n \in \mathcal{K}$ is a decreasing sequence, $K_n \downarrow K$,

$$f(K) = \lim_{n\to\infty} f(K_n) .$$

A set $A \subset E$ is called f-$\mathcal{K}$-capacitable iff

$$f(A) = \sup\{f(K) : K \in \mathcal{K}; K \subset A\} .$$

In general we don't have an analogue of 9.7

since there we critically used compactness. However there is an analogue of 9.3. To describe it, we require the notion of an $\mathcal{K}$-Souslin set.

The definition of $\mathcal{K}$-Souslin sets proceeds as follows. Let S denote the family of finite sequences $(n_1, \ldots, n_k)$ for $n_i \in \mathbb{N}$, the positive integers. Consider a family $\mathfrak{F} = \{X_\lambda\}$ for $\lambda \in S$ and for each $\sigma \in \mathbb{N}^{\mathbb{N}}$, $\sigma = (n_1, n_2, \ldots)$, let

$$X_\sigma = \bigcap_{i=1}^{\infty} X_{n_i \ldots n_i}$$

and

$$\mathcal{A}_{\mathfrak{F}} = \bigcup \{X_\sigma : \sigma \in \mathbb{N}^{\mathbb{N}}\} .$$

This procedure is called an $\mathcal{A}$-operation.

In the case above the set of all $\mathcal{K}$-<u>Souslin</u> sets are defined as the set of all $\mathcal{A}_{\mathfrak{F}}$ as $\mathfrak{F} = \{X_\lambda\}$ ranges over all indexing's; that is, maps $\mathfrak{F} : S \to \mathcal{K}$ with $X_\lambda \in \mathcal{K}$.

The $\mathcal{K}$-Souslin sets are stable under countable unions and intersections and so contain the $\mathcal{K}$-Borel sets; that is the monotone class generated by $\mathcal{K}$.

Proofs and further details may be found in Sierpinski, [1] <u>General Topology</u>. We merely state the following:

9.15 THEOREM. *Under the hypotheses of* 9.14, *every Souslin set is* f-$\mathcal{K}$-*capacitable*.

9.16 THEOREM. *In a* σ-*compact Hausdorff space, the* $\mathcal{K}$-*Souslin sets coincide with the* $\mathcal{K}$-*analytic sets*.

PROBLEMS FOR §9

9.1 *The space of compact subsets*

Show that $\mathcal{K}$ in 9.13 is complete.

9.2 *A capacity in the plane*

Let $X \subset \mathbb{R}^2$ be a set. For $0 < \alpha \leq \pi$, let Δ_α be the image of the x-axis under a counter-clockwise rotation of angle α and let $f(X,\alpha)$ denote the outer Lebesgue measure of the orthogonal projection of X on Δ_α.

(i) For X compact, show that the map $\alpha \mapsto f(X,\alpha)$ is upper semi-continuous. Is compactness necessary?

(ii) Let $f(X) = \int_0^\pi f(X,\alpha)d\alpha$. Show that f is a regular capacity.

(iii) What can be deduced concerning the f-capacitability of X when it is a Borel set?

§10 ORDERED VECTOR SPACES

Vector spaces are often equipped with an order relation compatible with the vector space structure. The example one should keep in mind is $\mathcal{K}(E,\mathbb{R})$ with the order $f \leq g$ iff $f(x) \leq g(x)$ for all $x \in E$.

The main results of this section are the Riesz decomposition lemma (10.5), which is fundamental in proving that certain functions are linear, and the description of the bounded linear forms on a space (10.9).

10.1 <u>DEFINITION</u>. Let E be a real vector space, and $\leq$ a partial ordering on E. (See section 1). We say E is an <u>ordered vector space</u> iff

(i) $(x \leq y \text{ and } z \in E) \Rightarrow (x + z \leq y + z)$

and (ii) $(x \leq y \text{ and } \alpha \in \mathbb{R},\ \alpha \geq 0) \Rightarrow (\alpha x \leq \alpha y)$.

If E is an ordered vector space the <u>positive cone</u> E^+ is defined by $E^+ = \{x \in E: x \geq 0\}$ and the <u>negative cone</u> is defined by $E^- =$ $E^- = \{x \in E: x \leq 0\}$.

Clearly $x,y \in E^+$, $\alpha \geq 0$ implies $x + y \in E^+$ and $\alpha x \in E^+$, by (i) and (ii) respectively. Also note that $E^+ \cap E^- = \{0\}$ and $x \in E^+$ iff $-x \in E^-$, that is, $E^- = -E^+$. [<u>Proof</u>: $x \geq 0$ implies $x - x \geq -x$, or $-x \leq 0$.] The set E^+ <u>is convex:</u> $x,y \in E^+$ implies $\lambda x + (1-\lambda) y \in E^+$ for all $0 \leq \lambda \leq 1$. Since the order is only partial, an element $x \in E$ need not be positive or negative.

For example, let S be a set and E a vector space of real functions $f: X \to \mathbb{R}$ with $f \leq g$ iff $f(x) \leq g(x)$ for all $x \in S$. Then E is an ordered vector space. Also on $\mathbb{R}^2$, the relation $(x_1,x_2) \leq (y_1,y_2)$ iff $|y_1 - x_1| \leq y_2 - x_2$ defines an order (see figure 10.1)

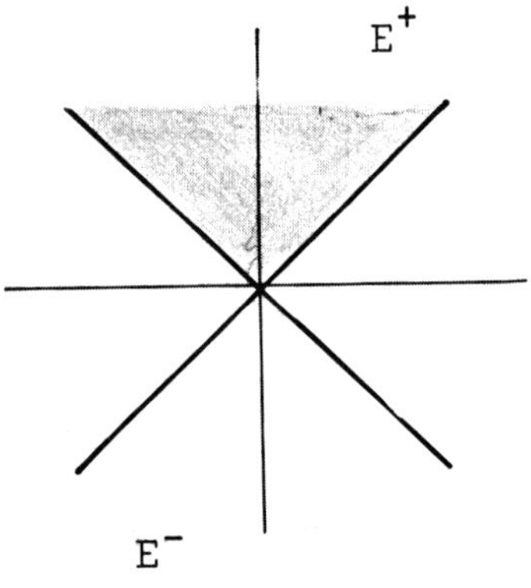

Figure 10.1

10.2 <u>PROPOSITION</u>. <u>Let</u> E <u>be a real vector space and</u> E^+ <u>a subset of</u> E <u>satisfying</u>:

(i) E^+ <u>is stable under addition and multi-</u>

<u>plication by positive scalars</u> (i.e. E^+ <u>is a convex cone</u>)

<u>and</u> (ii) $E^+ \cap (-E^+) = \{0\}$. (<u>That is</u>, E^+ <u>is a proper, pointed cone</u>);

<u>then there is a unique partial order on</u> E <u>making it into an ordered vector space for which</u> E^+ <u>is the positive cone, namely</u> $(x \geq y) \Longleftrightarrow (x - y \in E^+)$.

The proof is a straightforward check.

For example if I is any set we can define $(\mathbb{R}^I)^+ = (\mathbb{R}^+)^I$ making $\mathbb{R}^I$ into an ordered vector space.

10.3 <u>DEFINITION</u>. Let E be an ordered vector space. Then E is called a <u>lattice vector space</u> (or a vector lattice) iff for $x,y \in E$ there exists a least upper bound of the set $\{x,y\}$, denoted by $\sup(x,y) = x \vee y$. Also, we put $x \wedge y = -\sup(-x,-y)$; $x^+ = x \vee 0$, $x^- = (-x)^+ = (-x) \vee 0 = -(x \wedge 0)$ and, $|x| = x^+ + x^-$.

Clearly $x \wedge y = \inf(x,y)$. Also note that

$$(x + a) \vee (y + a) = a + (x \vee y)$$

and $$(\lambda x) \vee (\lambda y) = \lambda(x \vee y) \quad \text{for } \lambda \geq 0,$$

with similar formulae for inf's.

10.4 PROPOSITION. *Let* E *be a vector lattice. Then* (i) $x = x^+ - x^-$, *which implies* $E = E^+ + E^-$.

(ii) $x \vee y = (x+y)/2 + |y-x|/2$.

(iii) $x \wedge y = (x+y)/2 - |y-x|/2$.

(iv) $x^+ \wedge x^- = 0$; *that is*, x^+ *and* x^- *are disjoint*.

(v) $|x| = x \vee (-x)$.

(vi) $|x+y| \leq |x| + |y|$.

The proof is an easy verification. For example to prove (i), note that $x + x^- = x + (-x) \vee 0 = 0 \vee x = x^+$. For (iv), we use (iii) to give $x^+ \wedge x^- = |x|/2 - |x^+ - x^-|/2$ and the fact that $x^+ - x^- = x$. Finally, (v) follows from (ii), and the fact that $|x| = |-x|$. For (vi) note that $x^+ + y^+ \geq x + y$ and so $x^+ + y^+ \geq (x+y)^+$. Similarl $x^- + y^- \geq (x+y)^-$ from which the result follows. □

Notice that for E to be a lattice vector space it is enough that $x \vee 0$ exists for all $x \in E$.

Sometimes a lattice vector space is called a *Riesz space*.

10.5 RIESZ DECOMPOSITION LEMMA. *Let* E *be a vector lattice and* $x, y_1, y_2 \geq 0$ *with* $x \leq y_1 + y_2$. *Then there exist* $x_1, x_2 \geq 0$ *such that*

$$x = x_1 + x_2, \; x_1 \leq y_1, \text{ and } \; x_2 \leq y_2.$$

Proof. Let $x_1 = x \wedge y_1$ and $x_2 = x - x_1$. Clearly $0 \leq x_1 \leq y_1$ and $0 \leq x_2$. Also

$$x \leq (x + y_2) \wedge (y_1 + y_2) = y_2 + x \wedge y_1$$

so that

$$x_2 = x - x \wedge y_1 \leq y_2 . \; \square$$

If we write $[x,y] = \{a \in E: x \leq a \leq y\}$ then 10.5 states that

$$[0,y_1] + [0,y_2] = [0,y_1 + y_2]$$

for $y_1, y_2 \geq 0$.

An ordered vector space satisfying 10.5 need not be a vector lattice, and it will be useful to make the distinction.

10.6 COROLLARY. *Let* E *be an ordered vector space with the Riesz decomposition property* (10.5) *and let* $x_1,\dots,x_n \in E^+$, *and* $y_1,\dots,y_m \in E^+$ *with* $\Sigma\{x_i\} = \Sigma\{y_j\}$. *Then there exist* $z_{ij} \in E^+$, $i = 1,\dots,n,\ j = 1,\dots,m$ *such that*

$$x_i = \sum_{j=1}^{m} z_{ij} \quad \text{and} \quad y_j = \sum_{i=1}^{n} z_{ij}.$$

This follows at once from 10.5 by induction on n, m.

10.7 DEFINITION. Let E be an ordered vector space. We say E is a *weakly Riesz space* (or *weak vector lattice*) iff the decomposition property 10.5 holds and if E^+ generates E (i.e. $E = E^+ + E^-$).

Let E be an ordered vector space and

$$E^* = \{L\colon E \to \mathbf{R}\colon L \text{ is linear}\}$$

with the order induced by

$$(E^*)^+ = \{L \in E^*\colon x \geq 0 \text{ implies } L(x) \geq 0\}.$$

An element $L \in E^*$ is called *bounded* iff for $a,b \in E$, $a \leq b$, $L([a,b])$ is a bounded set in $\mathbf{R}$. The set of bounded elements of E^* is denoted B(E)

10.8 THEOREM. *Let* E *be a weakly Riesz space.*

Then $L \in E^*$ _is bounded iff there exist_ $L_1, L_2 \in (E^*)^+$ _such that_ $L = L_1 - L_2$.

Half of the theorem is easy. If $L \in (E^*)^+$ then $a \leq b$ implies $L(a) \leq L(b)$ so that $L([a,b]) \subset [L(a), L(b)]$. Therefore, if $L_1, L_2 \in (E^*)^+$, then $L = L_1 - L_2$ is bounded.

Because of 10.4, theorem 10.8 is a special case of the following:

10.9 _THEOREM_. _Let_ E _be a weakly Riesz space_. _Then_ $B(E) \subset E^*$ _is a vector lattice_. _In fact_, $B(E)$ _is a complete lattice_ (_every set_ $S \subset B(E)$ _with an upper bound has a least upper bound_ $\alpha \in B(E)$.)

Proof. Given 0 , $L \in B(E)$ it is sufficient to show $L \vee 0$ exists to prove that $B(E)$ is a vector lattice. Define $N \in E^*$ by

$$N(a) = \sup\{L(x): 0 \leq x \leq a\}$$

for $a \in E^+$ (and extend to all of E by linearity). We must show that N is linear on E^+ .

Clearly $N(\alpha x) = \alpha N(x)$ for $\alpha \in \mathbb{R}^+$, $x \in E^+$.

Secondly suppose that $0 \le x \le a+b$. Then there exist u_1, u_2 such that $x = u_1 + u_2$ and $0 \le u_1 \le a$, $0 \le u_2 \le b$. Thus

$$N(a+b) = \sup\{L(u_1+u_2)\colon 0 \le u_1 < a;\ 0 \le u_2 \le b\}$$
$$= N(a) + N(b) .$$

Now $N \ge 0$ and $N \ge L$ by definition. Also, N is the least upper bound of L and 0 . Indeed $M \ge 0$, $M \ge L$ implies that for all $0 \le x \le a$, $L(x) \le M(x) \le M(a)$ so that $M(a) \ge N(a)$. Since $N \ge 0$, $N \in B(E)$.

Now observe that $B(E)$ is complete iff every $X \subset B(E)^+$ has a greatest lower bound. (For $Y \subset B(E)$ with $u \ge Y$ consider the set $V = \{u - y\colon y \in Y\}$) . For $X \subset B(E)^+$, let X' denote the set of greatest lower bounds of finite subsets of X . Clearly $\inf X' = \inf X$ if it exists. Note that the set $\mathfrak{F}$ of finite subsets of X forms a directed set. For each $F \in \mathfrak{F}$ let $L_F = \inf\{L\colon L \in F\}$. For each $x \in E^+$, $L_F(x)$ forms a decreasing net in $\mathbb{R}^+$, and so converges to, say $L(x)$. ($L_F(x) \le L_{F_1}(x)$ if $F \supset F_1$) . Clearly L (extended by linearity) is the required greatest lower bound. □

10.10 COROLLARY. *Let* E *be a weakly Riesz space and* $L, M \in B(E)$ *. Then we have, for all* $a \geq 0$,

(i) $L \vee M(a) = \sup\{L(x) + M(y) : x \geq 0, y \geq 0, x + y = a\}$.

(ii) $L \wedge M(a) = \inf\{L(x) + M(y) : x \geq 0, y \geq 0, x + y = a\}$.

(iii) $L^+(a) = \sup\{L(x) : 0 \leq x \leq a\}$.

(iv) $|L|(a) = \sup\{L(x) : |x| \leq a\}$.

(v) $|L(y)| \leq |L|(|y|)$ for all $y \in E$.

Proof. From the above proof, $L \vee 0(a) = \sup\{L(x) : 0 \leq x \leq a\}$. Then $L \vee M = (L - M) \vee 0 + M$ so that

$$L \vee M(a) = \sup\{L(x) - M(x) : 0 \leq x \leq a\} + M(a)$$
$$= \sup\{L(x) + M(a - x) : 0 \leq x \leq a\}$$

which gives (i). Then (ii) follows at once using the fact that $L \wedge M = -(-L \vee -M)$. For (iii), recall that $L^+ = L \vee 0$. To prove (iv), note that $|L| = L \vee (-L)$ (10.4(v)) and so by (i),

$$|L|(a) = \sup\{L(x - y) : x \geq 0, y \geq 0, x + y = a\}$$
$$= \sup\{L(z) : |z| \leq a\} ,$$

(since $\{x - y : x \geq 0, y \geq 0, x + y = a\} = \{z : |z| \leq a\}$ as is easily seen by writing $z = z^+ - z^-$, and using 10.4(vi).)

Finally, (v) follows at once from (iv). □

10.11 COROLLARY. Let E be a weakly Riesz space and $L, M \in B(E)^+ = (E^*)^+$. Then L and M are disjoint ($L \wedge M = 0$) iff for all $a > 0$ and $\varepsilon > 0$, $a \in E$, $\varepsilon \in \mathbb{R}$, there exists $x, y \in E$ such that $L(x) + M(y) < \varepsilon$ and $x + y = a$.

Proof. Immediate from 10.10 (ii). □

Remark. Theorems 10.8 and 10.9 generalize by replacing $\mathbb{R}$ by a complete lattice E_1, so that E^* becomes $L(E, E_1)$ the set of linear maps $L: E \to E_1$ and $B(E, E_1)$ is the set of bounded linear maps. (A map is bounded iff it takes bounded sets to bounded sets; $X \subset E$ is bounded iff $X \subset [a,b]$ for some $a,b \in E$). The proofs are unchanged.

10.12 EXAMPLES. (i) Let X be a topological space and $\mathcal{C}(X,\mathbb{R})$ the vector space of continuous maps $f: X \to \mathbb{R}$. This is a lattice vector space with the order defined by $f \geq 0$ iff $f(x) \geq 0$ for all $x \in X$. In general this vector lattice is not complete. (See exercise 10.1 (iii)).

The subspace $\mathcal{K}(X,\mathbb{R})$ of continuous functions with compact support is also a vector lattice. If X is Euclidean space, then the r times continuously differentiable functions, denoted by $\mathcal{C}^r(X,\mathbb{R})$, $r \geq 1$, form an ordered vector space but not a lattice (in the usual order).

(ii) (conical measures). Let X be a normed vector space (or a topological vector space) and X' its topological dual, $X' = \{L: X \to \mathbb{R}: L$ is linear and continuous$\}$. Let $\mathfrak{F}$ denote the set of finite subsets of X' and for $F \in \mathfrak{F}$, $x \in X$, let $(\sup F)(x) = \sup\{L(x): L \in F\}$. Now let

$$h(X) = \{: X \to \mathbb{R}: \text{there exists } F_1, F_2 \in \mathfrak{F} \text{ such that } f(x) = (\sup F_1)(x) - (\sup F_2)(x) \text{ for all } x\}.$$

It is easy to check that $h(X)$ is a vector space. It is ordered by $f \geq 0$ iff $f(x) \geq 0$ for all $x \in X$. It is, in fact, a vector lattice (but not a complete lattice).

$((\sup F_1(x) - \sup F_2(x)) \vee 0 = \{\sup F_1(x) \vee \sup F_2(x)\} - \sup F_2(x)$, and $\sup F_1(x) \vee \sup F_2(x) = \sup F_1 \cup F_2(x$

Elements of $[h(X)^*]^+ \subset B(h(X))$ are called conical measures. (A complete lattice by 10.9).

We define $a(X)$ similar to $h(X)$ except that we replace X' by affine forms; that is, continuous linear forms plus constants. Then elements of $[a(X)^*]^+$ are the affine measures. (These contain the <u>cylinder measures</u>). We shall return to these examples later (chapter 9).

(iii) Let A be a normed space and $X \subset A$ be convex and compact. (X is called <u>convex</u> iff $x_1, x_2 \in X$, $0 \leq \lambda \leq 1$ implies $\lambda x_1 + (1-\lambda)x_2 \in X$.) A map $f: X \to \mathbb{R}$ is <u>convex</u> iff $f(\lambda x_1 + (1-\lambda)x_2) \leq \lambda f(x_1) + (1-\lambda) f(x_2)$, for all $x_1, x_2 \in X$, and $0 \leq \lambda \leq 1$.

Let $E = \{f \in \mathcal{C}(X,\mathbb{R})$: there exist $f_1, f_2: X \to \mathbb{R}$, continuous and convex with $f = f_1 - f_2\}$.

Then E is a lattice vector space with $f \geq 0$ iff $f(x) \geq 0$ for all x. By Stone Weierstrass (lattice version), E is dense in $\mathcal{C}(X,\mathbb{R})$. ($E$ separates points by the Hahn-Banach theorem (proven later) and linear maps are convex.)

(iv) (<u>Harmonic functions</u>). Let Ω be an open connected subset of $\mathbb{R}^n$. A C^2 map $\varphi: \Omega \to \mathbb{R}$ is <u>harmonic</u> iff $\Sigma\{\partial^2\varphi/\partial x_1^2\} = 0$ (Laplacian). Harmonic functions are in fact analytic. Let E denote the space of harmonic functions $\varphi: \Omega \to \mathbb{R}$ with order $\varphi \geq 0$ iff $\varphi(x) \geq 0$ for all x. Clearly E is an ordered vector space. It is in fact a lattice although this is not obvious. This is not too hard to see in case $n = 2$, Ω is the unit ball, and $F \subset E$ is the set of harmonic functions with continuous boundary values. Then the classical Poisson integral provides a 1-1 correspondence between continuous functions on the boundary and harmonic functions on the interior with these boundary values. Since the continuous functions on the boundary form a lattice, so do the harmonic functions. In general, however, $f_1 \wedge f_2(x) \neq \inf\{f_1(x), f_2(x)\}$; see figure 10.2.

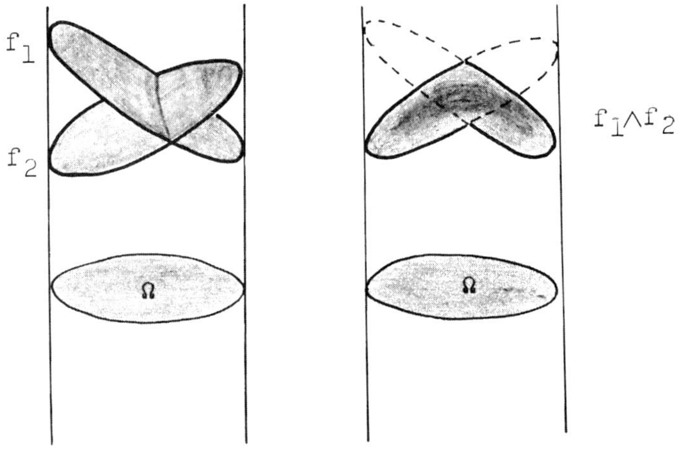

Figure 10.2

PROBLEMS FOR §10

10.1 Miscellaneous exercises on lattice vector

spaces.

(i) Let E be an ordered vector space with a function $|\cdot|: E \to E^+$ satisfying (a) $|x| \geq x$, (b) $|x| + x \geq 0$ and (c) $(y \geq 0, y \geq x) \Longrightarrow 2y \geq |x| + x$. Show that E is a lattice.

(ii) If E is a lattice vector space, $x, y \in E^+$, $x \wedge y = 0$ and $z = x - y$, then $z^+ = x$ and $z^- = y$.

(iii) Show that the smallest complete lattice vector space containing $\mathcal{C}(E,\mathbb{R})$ is $\mathbb{R}^E$, where E is a locally compact Hausdorff space.

10.2 A lattice of linear forms.

Let L be a positive linear form on a lattice vector space E and $0 < a \in E^+$. Define the following subset of $(E^*)^+$:

$$A_{L,a} = \{M \in (E^*)^+ : 0 \leq M \leq L \text{ and } (0 \leq x \leq a) \Rightarrow (M(x) = L(x)) \text{ and } (x \geq 0, x \wedge a = 0) \Rightarrow (M(x) = 0)\}$$

(i) $A_{L,a}$ has a greatest element L_a.

(ii) For $\lambda > 0$, $L_{\lambda a} = L_a$.

(iii) If $a \wedge b = 0$ then $L_{a+b} \leq L_a + L_b$.

(iv) Equality in (iii) can fail.

10.3 The sup of disjoint elements

(i) Let E be a vector lattice and $x_1, x_2 \in E$. We say that x_1 and x_2 are <u>disjoint</u> iff $|x_1| \wedge |x_2| = 0$. Show that x_1 and x_2 are disjoint iff $|x_1| \vee |x_2| = |x_1| + |x_2|$.

(ii) If x_1 and x_2 are disjoint, show that $(x_1 + x_2)^+ = x_1^+ + x_2^+$ and $|x_1 + x_2| = |x_1| + |x_2|$.

§11 INTEGRATION OF RADON MEASURES

We have already seen one approach to integration theory in §3. There is another approach, namely that of Radon measures. Although the positive Radon measures and regular Borel measures are in one to one correspondence by means of the Riesz representation theorem, it is useful to see the development from the Radon measure point of view. We shall be brief in this regard and we refer the reader to Bourbaki, [3] Integration, (XIII), for a more complete treatment. Therefore many proofs pertaining to the measure theoretic aspect of the work will be omitted.

The main results needed for later sections are the basic properties (11.2), and the definitions of norms, supports etc. (11.20-11.22).

11.1 DEFINITIONS. Let E be a locally compact Hausdorff space and $\mathcal{K}(E,\mathbb{R})$ the continuous functions $f: E \to \mathbb{R}$ with compact support. On $\mathcal{K}(E,\mathbb{R})$ put the order defined by $f \geq 0$ iff for all $x \in E$, $f(x) \geq 0$.

A Radon measure on E is a linear map $\mu : \mathcal{K}(E,\mathbb{R}) \to \mathbb{R}$ such that for any $K \subset E$ compact there exists a number M_K such that $f \in \mathcal{K}(E,\mathbf{R})$, $S(f) \subset K$, implies $|\mu(f)| \leq M_K\|f\|$. (Recall that $S(f)$ is the support of f and $\|f\| = \sup\{|f(x)|: x \in E\}$.) The collection of Radon measures on E is denoted $\mathcal{M}(E)$ and the set of positive Radon measures (with the order induced by $\mathcal{K}(E,\mathbf{R})^*)^+$ is denoted by $\mathcal{M}^+(E)$ (that is, $\mu \geq 0$ iff $f \in \mathcal{K}(E,\mathbf{R})^+$ implies $\mu(f) \geq 0$.

Thus $\mu \in \mathcal{K}(E,\mathbf{R})^*$ is a Radon measure iff for every compact $K \subset E$, the restriction of μ to $\mathcal{K}_K(E,\mathbf{R})$ is continuous in the uniform norm. Therefore, if we put the inductive topology on $\mathcal{K}(E,\mathbf{R})$ (see 5.2), $\mu \in \mathcal{K}(E,\mathbf{R})^*$ is a Radon measure iff μ is continuous. Thus, $\mathcal{M}(E) = \mathcal{K}(E,\mathbf{R})'$. This fact will be important later, for we can then apply general theorems about duals of topological vector spaces (see section 12).

11.2 THEOREM. Let E be a locally compact Hausdorff space.

(i) $\mathcal{K}(E,\mathbf{R})$ is a vector lattice,

(ii) $\mathcal{K}(E,\mathbf{R})^*)^+ = \mathcal{M}^+(E)$; that is, all positive linear forms on $\mathcal{K}(E,\mathbf{R})$ are Radon measures (*that is, continuous*),

(iii) $\mathcal{M}(E) = B(\mathcal{K}(E,\mathbf{R}))$ *in the notation of* 10.9,

(iv) $\mathcal{M}(E)$ *is a complete lattice*,

(v) *for* $\mu \in \mathcal{M}(E)$, *we can write* $\mu = \mu^+ - \mu^-$ *where* $\mu^+, \mu^- \geq 0$ *and* $\mu^+ \wedge \mu^- = 0$ (*disjoint*), *and* $|\mu| = \mu^+ + \mu^-$. (*Later, this will be shown to coincide with the classical Jordan-Hahn decomposition*).

Proof. If $f, g \in \mathcal{K}(E,\mathbf{R})$ then $f \vee g \in \mathcal{C}(E,\mathbf{R})$ and $S(f \vee g) \subset S(f) \cup S(g)$, so $f \vee g \in \mathcal{K}(E,\mathbf{R})$. This proves (i).

For (ii), let $K \subset E$ be compact and $\mu \geq 0$. Since E is locally compact Hausdorff there is a continuous map $\varphi: E \to [0,1]$ with compact support and $\varphi(K) = 1$. Let $M_K = 2\mu(\varphi)$. Then if $f \in \mathcal{K}(E,\mathbf{R})^+$ and $S(f) \subset K$,

$$0 \leq f = f\varphi \leq \|f\|\varphi \quad \text{which implies}$$

$$0 \leq \mu(f) \leq \mu(\|f\|\varphi) \leq M_K \|f\|/2 .$$

More generally we see that $|\mu(f)| \leq M_K \|f\|$ by writing $f = f^+ - f^-$.

Next we prove (iii). Let $f, g \in \mathcal{K}(E,\mathbf{R})$ and $\mu \in \mathcal{M}(E)$. We must show $\mu([f,g])$ is bounded.

Let $K = S(f) \cup S(g)$ and $f \leq h \leq g$. Then $S(h) \subset K$ and $|\mu(h)| \leq M_K \|h\| \leq M_K \sup\{\|f\|, \|g\|\}$. Thus, $\mathcal{M}(E) \subset B(\mathcal{K}(E,\mathbf{R}))$. By (ii), $\mathcal{M}^+(E) = B^+(\mathcal{K}(E,\mathbf{R}))$ and hence $\mathcal{M}(E) \supset B(\mathcal{K}(E,\mathbf{R}))$. Therefore we have equality.

Finally (iv) and (v) are immediate by 10.4 and 10.9. □

We now take up the problem of extending $\mu \geq 0$ from $\mathcal{K}(E,\mathbf{R})$ to a larger class of functions. To motivate the definition recall that regular Borel measures are continuous under increasing limits of continuous functions (or lower semi-continuous functions), by the monotone convergence theorem. (If the space is σ-compact, an arbitrary increasing limit may be reduced to a countable one.)

11.3 DEFINITION. Let E be locally compact Hausdorff, $\mu \in \mathcal{M}^+(E)$, and $\mathcal{J}^+ = \mathcal{J}^+(E)$ be the set of lower semi-continuous mappings $f\colon E \to [0,\infty]$. For $f \in \mathcal{J}^+$ we put $\mu^*(f) = \sup\{\mu(g)\colon g \in \mathcal{K}(E,\mathbf{R})^+ , g \leq f\}$.

11.4 LEMMA. (i) Let $f \in \mathcal{J}^+(E)$, then

$f = \sup\{g \in \mathcal{K}(E,\mathbb{R})^+ : g \leq f\}$,

(ii) *suppose* $\{f_i\}$ *is an increasing* (*or filtering increasing*) *net in* $\mathcal{K}^+(E,\mathbb{R})$ *and* $f = \sup(f_i) \in \mathcal{K}^+(E,\mathbb{R})$. *Then if* $\mu \in \mathcal{M}^+(E)$, $\mu(f) = \sup\{\mu(f_i)\}$.

Proof. (i) Let $x \in E$ and $f(x) > 0$. Let $0 < a < f(x)$. Choose a neighborhood U of x on which $f > a$. By Urysohn's lemma there exists $g \in \mathcal{K}^+(E,\mathbb{R})$ with support in U so that $g \leq f$ and $g(x) = a$. Hence $f(x) = \sup\{g(x) : g \in \mathcal{K}^+(E,\mathbb{R});\ g \leq f\}$ which proves (i).

For (ii), we have $\mu(f) \geq \sup \mu(f_i)$. Since $f, f_i \in \mathcal{K}^+(E,\mathbb{R})$ and the limit is increasing (or filtering increasing), it is uniform (Dini's theorem). Hence from the estimate $|\mu(f - f_i)| \leq M_K \|f - f_i\|$ we have (ii). □

Basic properties of μ^* are the following:

11.5 *THEOREM*. *Let* E *be a locally compact Hausdorff space and* $\mu \in \mathcal{M}^+(E)$. *Then*

(i) if $f \in \mathcal{J}^+(E)$, $a \geq 0$, $\mu^*(af) = a\mu^*(f)$,

(ii) if $f, g \in \mathcal{J}^+(E)$, $f \leq g$ *then* $\mu^*(f) \leq \mu^*(g)$,

(iii) if $f, g \in \mathcal{J}^+(E)$ *then*

$\mu^*(f+g) = \mu^*(f) + \mu^*(g)$,

(iv) *if* $\{f_i\}$ *is an increasing net in* $\mathcal{J}^+(E)$ *and* $f = \sup\{f_i\}$ *then* $f \in \mathcal{J}^+(E)$ *and*

$$\mu^*(f) = \sup\{\mu^*(f_i)\}$$

(v) *if* $\{g_i\}$ *is any net in* $\mathcal{J}^+(E)$ *then* $\Sigma g_i \in \mathcal{J}^+(E)$ *and*

$$\mu^*(\Sigma g_i) = \Sigma\mu^*(g_i) .$$

(*where* $(\Sigma g_i)(x) = \Sigma g_i(x)$) .

Proof. (i) and (ii) are clear. For (iii) note that

$$\begin{aligned}\mu^*(f+g) &= \sup\{\mu(h)\colon h \in \mathcal{K}(E,\mathbf{R})^+,\ h \le f+g\} \\ &= \sup\{\mu(f_1+g_1)\colon f_1,g_1 \in \mathcal{K}^+(E,\mathbf{R}), \\ &\qquad f_1 \le f\ ,\ g_1 \le g\}\end{aligned}$$

by the Riesz decomposition property.

To prove (iv), we have for each $\lambda \in \mathbf{R}\cup\{+\infty\}$

$$\{x \in E\colon f(x) \le \lambda\} = \bigcap_i \{x \in E\colon f_i(x) \le \lambda\}$$

so that $f \in \mathcal{J}^+$. Also, $\mu^*(f_i) \le \mu^*(f)$ so that $\sup\{\mu^*(f_i)\} \le \mu^*(f)$. Let $\Phi = \cup\Phi_i$ where $\Phi_i = \{g \in \mathcal{K}^+ : g \le f_i\}$. For any $h \in \mathcal{K}^+(E,\mathbf{R})$ with $h \le f$ we have

$$\mu(h) = \sup\{\mu(g \wedge h): g \in \Phi\}$$
$$\leq \sup\{\mu(g): g \in \Phi\} = \sup\{\mu^*(f_i)\}$$

by using 11.4. Hence $\mu^*(f) \leq \sup\{\mu^*(f_i)\}$ proving the assertion. (v) is a consequence of (iv) and (iii). □

Notice that the net in 11.5 (iv) need not be countable. Later, for non semi-continuous functions, countability will be necessary.

For $U \subset E$ open notice that the characteristic function 1_U is lower semi-continuous. Hence we make:

11.6 DEFINITION. Let E be a locally compact Hausdorff space and $\mu \in \mathcal{M}^+(E)$. For $U \subset E$ open let $\mu^*(U) = \mu^*(I_U)$.

A set $X \subset E$ is called negligible (or of measure zero) iff for any $\epsilon > 0$ there is an open set $U \supset X$ such that $\mu^*(U) < \epsilon$.

For any map $f: E \to [0,\infty]$, let

$$\mu^*(f) = \inf\{\mu^*(g): g \in \mathcal{I}^+(E);\ g \geq f\}$$

and for a set $X \subset E$, $\mu^*(X) = \mu^*(1_X)$.

11.7 PROPOSITION. (i) *if* $\{U_i\}$ *is a disjoint family of open sets then* $\mu^*(\bigcup U_i) = \Sigma\mu^*(U_i)$

(ii) $X \subset E$ *is negligible iff* $\mu^*(X) = 0$.

Proof. (i) follows at once from 11.5 (iv), and (ii) follows easily from the definitions. □

More basic properties of μ^* are:

11.8 THEOREM. *Let* E *be a locally compact Hausdorff space and* $\mu \in \mathcal{M}^+(E)$. *All functions* f *below are mappings*: $E \to [0,\infty]$

(i) *if* $f \leq g$, *then* $\mu^*(f) \leq \mu^*(g)$

(ii) *for* $a \geq 0$, $\mu^*(af) = a\mu^*(f)$

(iii) $\mu^*(f_1 + f_2) \leq \mu^*(f_1) + \mu^*(f_2)$

(iv) (*Monotone convergence theorem*). *If* f_n *is an increasing sequence and* $f(x) = \lim_{n\to\infty} f_n(x)$ *then* $\mu^*(f) = \lim_{n\to\infty} \mu^*(f_n)$

(v) (*Fatou's lemma*) *for a sequence* f_n

$\mu^*(\liminf_{n\to\infty} f_n) \leq \liminf_{n\to\infty} \mu\ (f_n)$

(vi) *for a sequence* g_n *we have*

$\mu^*(\Sigma g_n) \leq \Sigma\mu^*(g_n)$.

The proof of this theorem (except (iv)) is

straightforward. The techniques are quite analogous to those the reader is familiar with in measure theory. For details, see Bourbaki, Livre VI, Ch III, p. 110-112.

11.9 <u>COROLLARY</u>. <u>Let</u> $f: E \to [0,\infty]$ <u>be a mapping</u>. <u>Then</u> $\mu^*(f) = 0$ <u>iff</u> $f = 0$ a.e. ($\{x: f(x) > 0\}$ <u>is</u> μ-<u>negligible</u>).

<u>Proof</u>. If $\mu^*(f) = 0$ and $A = \{x \in E: f(x) > 0\}$, then $\infty\cdot 1_A \geq 1_A$ and $\mu(\infty\cdot 1_A) = \lim_{n\to\infty} \mu^*(nf) = 0$ by the monotone convergence theorem. Therefore, $\mu(1_A) = 0$ and hence $f = 0$ a.e.. Conversely, if $\mu(1_A) = 0$, then $\mu(\infty\cdot 1_A) = \lim_{n\to\infty} \mu(n\cdot 1_A) = 0$ and since $0 \leq \mu^*(f) \leq \mu(\infty\cdot 1_A) = 0$, $\mu^*(f) = 0$. □

In our context, the L_p spaces are obtained as follows:

11.10 <u>DEFINITION</u>. Let E be a locally compact Hausdorff space, $\mu \in \mathcal{M}^+(E)$ and F a Banach space. For any map $f: E \to F$, put, for each p, $1 \leq p < \infty$,

$$N_p(f) = (\mu^*|f|^p)^{1/p} \in [0,\infty]$$

where $|f|(x) = \|f(x)\|$, and let

$$\mathfrak{F}_p(E,F,\mu) = \{f\colon E \to F\colon N_p(f) < \infty\} .$$

11.11 THEOREM. N_p *is a complete semi-norm on* $\mathfrak{F}_p(E,F,\mu)$.

Proof. (Recall that a semi-norm satisfies all the norm axioms except $\|f\| = 0$ implies $f = 0$.) That N_p is a semi-norm follows from the Minkowski inequalities: for $f,g\colon E \to [0,\infty]$,

$$\{\mu^*(f+g)^p\}^{1/p} \leq \{\mu^*(f^p)\}^{1/p} + \{\mu^*(g^p)\}^{1/p} .$$

This follows from a standard convexity lemma; see Bourbaki XIII, p. 12.

For the rest of the proof, we use the following whose proof is left to the reader:

11.12 LEMMA. *Let* X *be a semi-normed space. Then* X *is complete iff for every sequence* $\{x_n\}$ *with* $\Sigma\|x_n\| < \infty$, Σx_n *converges* (*resp. iff for every sequence* $\{x_n\}$ *with* $\|x_n\| \leq 1/2^n$, Σx_n *converges*).

For the theorem, let $f_n \in \mathfrak{F}_p(E,F,\mu)$ with $\Sigma N_p(f_n) < \infty$. Let $g(x) = \Sigma\|f_n(x)\|$. Then using 11.8 we see that

$$N_p(g) \leq \Sigma N_p(f_n) < \infty .$$

(<u>Proof</u>. By 11.8 (iv), $\mu^*\{(\Sigma|f_n|)^p\}$ $= \lim_{n\to\infty} \mu\{(\sum_{n=1}^{N} |f_n|)^p\}$ now take the pth root and apply Minkowski's inequality.)

Next, $\{x: g(x) = \infty\}$ is negligible since $\mu^*(|g|^p) < \infty$ and therefore $\lim_{n\to\infty} \mu^*(|g|^p/n) = 0$ (see the proof of 11.9). For $g(x) = \infty$, let $f(x) = 0$, and for $g(x) < \infty$, let $f(x) = \Sigma f_n(x)$ which converges as F is complete. Since $\|f\| \leq g$, we have $f \in \mathfrak{F}_p(E,F,\mu)$. Therefore, since $\{x: g(x) = \infty\}$ is negligible, $N_p(f - \sum_{k=1}^{n} f_k)$ $= N_p(\sum_{k=n+1}^{\infty} f_k) \leq \sum_{k=n+1}^{\infty} N_p(f_k) \to 0$ as $n \to \infty$. □

11.13 <u>DEFINITION</u>. In addition to the hypotheses of 11.11, let $\mathcal{K}(E,F)$ denote the space of continuous maps $f: E \to F$ with compact support, so that $\mathcal{K}(E,F) \subset \mathfrak{F}_p(E,F,\mu)$. The norm closure of $\mathcal{K}(E,F)$ in $\mathfrak{F}_p(E,F,\mu)$ is denoted $\mathcal{L}_p(E,F,\mu)$. Let R denote the equivalence relation on $\mathcal{L}_p(E,F,\mu)$ defined by fRg iff $N_p(f - g) = 0$. Then put

$$L_p(E,F,\mu) = \mathcal{L}_p(E,F,\mu)/R .$$

Remark. $N_p(f-g) = 0$ iff $\mu^*(f-g) = 0$.

11.14 THEOREM. $L_p(E,F,\mu)$ *is a Banach space, with norm inherited from* N_p .

This follows at once from 11.11.

The *dominated convergence theorem* is as follows (the case $p = 1$ is the usual form).

11.15 THEOREM. (i) *Let* f_n *be a sequence in* $\mathcal{L}_p(E,F,\mu)$ *and* $\|f_n(x)\| \leq g(x)$ *almost everywhere for some* $g \in \mathcal{L}_p^+(E,R,\mu)$. *If* $f_n(x) \to f(x)$ *almost everywhere then* $f \in \mathcal{L}_p(E,F,\mu)$ *and* $\|f_n - f\| \to 0$.

(ii) Let $f_n \in \mathcal{L}_1(E,F,\mu)$ with $\sum_{n=1}^{\infty} f_n(x) \rightrightarrows f(x)$ *converging for almost all* $x \in E$. *Also suppose there exists* $g: E \to [0,\infty]$ *with* $\mu^*(g) < \infty$ *and* $\|\sum_{k=1}^{n} f_k(x)\| \leq g(x)$ *for each* n . *Then* $f \in \mathcal{L}_1(E,F,\mu)$ *and*

$$\mu^*(f) = \sum_{n=1}^{\infty} \mu^*(f_n) .$$

For the proofs, see Bourbaki [3], XIII, p. 140 150.

One other important fact is that $L_p(E,F,\mu)$

is a complete lattice (with $[f] \leq [g]$ iff $f \leq g$ a.e.), although $\mathcal{L}_p(E,F,\mu)$ is not. For example let $E = [0,1] \subset \mathbf{R}$, μ be Lebesgue measure, and $X \subset E$ a non-measurable set. For each $x \in X$ define $f_x: [0,1] \to \mathbf{R}$; by $f_x(y) = 1$ iff $x \neq y$; $f_x(x) = 0$. Then the infimum of $\{f_x: x \in X\}$ does not exist in $\mathcal{L}_p$, but does in L_p. Compare exercise 10.1 (iii). For the proof, see Bourbaki, VI, ch. IV, §3, no. 6.

Next we show how to define measurable sets and functions directly in terms of a Radon measure. (Written in terms of regular Borel measures, this is known as <u>Lusin's theorem</u>).

11.16 <u>DEFINITION</u>. Let E be a locally compact Hausdorff space and $\mu \in \mathcal{M}^+(E)$. A mapping $f: E \to \mathbf{R} \cup \{+\infty\}$ is called μ-<u>measurable</u> iff for every compact $K \subset E$ and $\varepsilon > 0$, there exists a compact $K_1 \subset E$ such that f is continuous at points of K_1 and $\mu(K \setminus K_1) < \varepsilon$.

A set $X \subset E$ is called μ-<u>measurable</u> iff its characteristic function is μ-measurable.

It follows easily that a set is measurable

iff for every compact K, the characteristic function of $X \cap K$ is μ-integrable (lies in $\mathcal{L}_1(E,\mathbf{R},\mu)$).

11.17 THEOREM. The μ-measurable sets form a σ-field and contain the Borel sets. Further, μ^* is a measure on the μ-measurable sets.

This follows fairly easily from the convergence theorems. See Bourbaki, pp. 177-186. For the classical criteria for measurability of functions, again see Bourbaki.

11.18 THEOREM. Let E be a locally compact Hausdorff space with countable base. Let $\mu \in \mathcal{M}^+(E)$ and ν be the regular Borel measure determined from the Riesz representation theorem. Then $\nu = \mu^*$ on the Borel sets. Also, $\mathcal{L}_1(E,\mathbf{R},\mu) = \mathcal{L}_1(\nu)$ and the measurable sets and functions coincide.

The method of proof is the following. Since μ^* and ν are both measures on the Borel sets it suffices to show that for U open, $\mu^*(U) = \nu(U)$. But there exists a sequence $f_n \in \mathcal{K}(E,\mathbf{R})$ such that $f_n \uparrow 1_U$ by normality and the fact that there is a sequence of compacts $K_n \subset U$ with $K_n \subset \text{int}(K_{n+1}$

and $K_n \uparrow U$. But $\mu^*(f_n) = \nu(f_n)$ and both converge to $\mu^*(U)$ and $\nu(U)$. The last part follows from the measurability criteria referred to above. □

Another link with Borel measures is the following:

11.19 <u>THEOREM</u>. *Let E be a locally compact Hausdorff space with countable base and* $\mu = \mu^+ - \mu^- \in \mathcal{M}(E)$. *Then* μ^+ *and* μ^- *may be identified with the Jordan decomposition of* μ . *In particular,* μ^+ *and* μ^- *are supported by disjoint sets* (μ^+ *is supported by* A *iff the complement has measure zero*).

<u>Proof</u>. Let $\mu^* = \nu^+ - \nu^-$ be the Jordan decomposition of μ^* . It suffices to show that, as Radon measures, we have $\nu^+ \wedge \nu^- = 0$. (See problem 10.1 (ii)). For this, we use the criterion of 10.11. Given $f \in \mathcal{K}(E,\mathbb{R})^+$, $\varepsilon > 0$ suppose that $S(f) \subset K$ and ν^+ is supported by A, ν^- by B . Choose open sets U, V such that $U \supset (A \cap K)$, $V \supset (B \cap K)$ and $|\mu|(U \setminus (A \cap K)) < \varepsilon'$ and $|\mu|(V \setminus (B \cap K)) < \varepsilon'$, where $\varepsilon' = \varepsilon/2\|f\|$. Choose functions g_1, g_2 such that $S(g_1) \subset U$, $S(g_2) \subset V$,

$g_1 + g_2 = 1$ on K , $0 \leq g_i \leq 1$ (partition of unity), and write $f = g_1 f + g_2 f$. Now $\nu^+(g_1 f) \leq \|f\|\varepsilon'$ and $\nu^-(g_2 f) \leq \|f\|\varepsilon'$, so that $\nu^+(g_1 f) + \nu^-(g_2 f) \leq 2\,\|f\|\varepsilon' = \varepsilon$. Hence by 10.11, ν^+ and ν^- are disjoint. □

Note that this proof does not use uniqueness of the Jordan decomposition but rather that a disjoint decomposition in a lattice is unique.

We return now to the setting of Radon measures and introduce some definitions which will be important later.

11.20 <u>DEFINITIONS</u>. Let E be a locally compact Hausdorff space and $\mathcal{M}(E)$ the space of Radon measures on E .

(i) For $\mu \in \mathcal{M}(E)$, define

$$\begin{aligned} \|\mu\| &= \sup\{|\mu(f)| : f \in \mathcal{K}(E,\mathbf{R});\ \|f\| \leq 1\} \\ &= \sup\{\ \mu(f)\ : f \in \mathcal{K}(E,\mathbf{R});\ \|f\| \leq 1\} \end{aligned}$$

(ii) A Radon measure $\mu \in \mathcal{M}(E)$ is said to be <u>zero</u> on an open set U iff $f \in \mathcal{K}(E,\mathbf{R})$, $S(f) \subset U$ implies $\mu(f) = 0$. The <u>support</u> of μ is defined by

$$S(\mu) = E \setminus (\bigcup\{\omega : \omega \text{ is open in } E \text{ and } \mu \text{ is zero on } \omega\}).$$

Further, for $\mu \geq 0$, $X \subset E$, we say that μ is <u>supported</u> (<u>carried</u>) by X iff $\mu^*(E \setminus X) = 0$.

It is important that the terms "support" and "supported by" not be confused. In our later work the distinction will be crucial.

11.21 <u>PROPOSITION</u>. *Let E be a locally compact Hausdorff space and $\mu \in \mathcal{M}(E)$. Then*

(i) μ *is zero on* $E \setminus S(\mu)$,

(ii) *if* $f \in \mathcal{K}(E,\mathbf{R})$ *and* $f = 0$ *on* $S(\mu)$ *then* $\mu(f) = 0$.

(iii) *For* $\mu_1, \mu_2 \in \mathcal{M}(E)$, $S(\mu_1 + \mu_2) \subset S(\mu_1) \cup S(\mu_2)$ *with equality if* $\mu_1 \geq 0$ *and* $\mu_2 \geq 0$.

(iv) *If* $\mu \geq 0$ *then* μ *is carried by* $S(\mu)$.

(v) *If* $\mu \geq 0$, *then* μ *is carried by* $X \subset E$ *iff for all* $f: E \to [0,\infty]$ *with* $f(X) = \{0\}$, *we have* $\mu(f) = 0$.

<u>Proof</u>. (i) Suppose $f \in \mathcal{K}(E,\mathbf{R})$ and $S(f) \subset E \setminus S(\mu)$. We must show that $\mu(f) = 0$. Let K be a compact neighborhood of $S(f)$, $K \subset S(f) \subset E \setminus S(\mu)$, and

let $\{g_i\}$ be a partition of unity on K subordinate to a finite covering of K by open sets on which $\mu = 0$. Then $f = \Sigma g_i f$ and $\mu(f) = \Sigma\mu(g_i f) = 0$ since each $\mu(g_i f) = 0$.

(ii) Let U be a neighborhood of $S(\mu)$ with $c\ell(U)$ compact. Choose $g\colon E \to [0,1]$ continuous, $S(g) \subset U$ and $g = 1$ on a neighborhood of $S(\mu)$ (Urysohn's lemma). Then by (i), $\mu(f\cdot(1-g)) = 0$. Thus $\mu(f) = \mu(fg)$ for any g. But by continuity of f, for $\varepsilon > 0$ there is a neighborhood U such that $|f| < \varepsilon$ on U. Thus $|\mu(fg)| \leq M|f| \leq M\varepsilon$, ($M$ is the constant for $K = S(f)$). Hence $\mu(f) = 0$.

(iii) That $S(\mu_1 + \mu_2) \subset S(\mu_1) \cup S(\mu_2)$ follows at once from the definition by taking complements. If $\mu_1, \mu_2 \geq 0$, we must show that if $\mu_1 + \mu_2 = 0$ on an open set ω, then μ_1 and $\mu_2 = 0$ on ω. If not, there exists $f \in \mathcal{K}(E,\mathbb{R})^+$, $S(f) \subset \omega$ such that, say $\mu_1(f) \neq 0$ (we can choose $f \geq 0$ because if μ_1 vanishes on $\mathcal{K}^+$, it vanishe on $\mathcal{K}$). Then $(\mu_1 + \mu_2)(f) \geq \mu_1(f) > 0$ a contradiction.

(iv) We must show that $\mu(E \setminus S(\mu)) = 0$. By definition 11.3, this means that for all $f \in \mathcal{K}^+$,

$0 \le f \le 1$ and $f = 0$ on $S(\mu)$, then $\mu(f) = 0$. But this follows by (ii).

(v) First, suppose μ is carried by X. Let g be the characteristic function on $E \setminus X$ and $h = \lim_{n\to\infty} ng$. Now $\mu(g) = 0$, so by the monotone convergence theorem 1.8, $\mu(h) = 0$. For a given $f\colon E \to [0,\infty]$ zero on X, $\mu(f) \le \mu(h)$ so that $\mu(f) = 0$. Choosing $f = g$, the converse is obvious. □

In the above proof, notice that one can also deduce directly (i) and (ii) from (v).

Perhaps an example will help clarify the distinction between "support" and "supported by". Namely, on the line, consider the measure consisting of Lebesgue measure on $[0,1]$. The support is $[0,1]$, but the measure is supported by the smaller set $]0,1[$.

Regarding the norm of a measure, note that if μ is bounded, $\|\mu\| < \infty$, we can regard μ as a continuous linear form on $\mathcal{C}_b(E,\mathbf{R})$ with the sup norm and $\|\mu\|$ is just its norm in the Banach space $\mathcal{C}_b(E,R)'$, where $\mathcal{C}_b(E,\mathbf{R})$ denotes the bounded continuous functions. See problem 12.7.

11.22 PROPOSITION. Let E be a locally compact Hausdorff space and $\mu \in \mathcal{M}(E)$. Then $\|\mu\| = |\mu|(1)$ and for any $f \in \mathcal{K}(E,\mathbf{R})$, $|\mu(f)| \leq |\mu|(|f|)$.

Proof. From 10.10 (iv) we have

$$|\mu|(f) = \sup\{\mu(g) : \|g\| \leq f\}$$

and hence $|\mu|(1) = \sup\{\mu(g) : \|g\| \leq 1\}$ from which the assertion is clear. The last part follows from 10.10 (v). □

PROBLEMS FOR §11

11.1 Diffuse measures.

Let $X \subset \mathbf{R}$ and suppose that for any sequence (r_n) of numbers, $r_n > 0$ there exists a sequence (A_n) of open intervals of $\mathbf{R}$ such that $X \subset \bigcup\{A_n\}$ and $|A_n| \leq r_n$ where $|\cdot|$ denotes length.

Let $\mu \in \mathcal{M}^+(\mathbf{R})$ be such that $\mu(\{x\}) = 0$ for each $x \in \mathbf{R}$ (μ is called diffuse or non-atomic). Then show that $\mu(X) = 0$ when 1_X is locally integrable.

11.2 A Lexicographically ordered space.

Let $E = [0,1] \times \{-1,0,1\}$ with the lexicographic order:

$$(x,i) \leq (x',i') \text{ iff } (x < x')$$
$$\text{or } (x = x' \text{ and } i \leq i')$$

Let $x_- = (x,-1)$, $x_0 = (x,0)$ and $x_+ = (x,1)$ for $x \in [0,1]$ and $E_- = \{0,1\} \times \{-1\}$, $E_0 = [0,1] \times \{0\}$, $E_+ = [0,1] \times \{1\}$. Give E the order topology.

(i) Show that the map $\varphi: E \to [0,1]$ defined by $\varphi((x,i)) = x$ is increasing and continuous.

(ii) Every compact set in E_- is well ordered; deduce that it is finite or denumerable.

(iii) Let $E' = E \setminus \{0_-, 1_+\}$. Find a canonical isomorphism between the continuous maps $\mathcal{C}(E';\mathbf{R})$ and the functions $f: [0,1] \to \mathbf{R}$ which at each point have a finite limit on the left and right.

(iv) E is compact.

(v) Using the isomorphism of (iii) define on E a positive Radon measure μ such that its image by φ is Lebesgue measure on $[0,1]$; is such a measure unique?

(vi) Show that for the measure μ of (v), E_0 is measurable and of 0-measure. Are E_- and E_+ measurable?

11.3 Representation of L_p functions

Let E be a locally compact Hausdorff space, $\mu \in \mathcal{M}^+(E)$ and $1 \leq p < \infty$. Show that every $f \in L_p(E,\mu)$ is almost everywhere equal to a function $g_1 - g_2$ with g_1, g_2 lower semicontinuous mappings of E into $[0, +\infty]$.

11.4 The limit of L_p as $p \to 0$

Let (E,μ) be a measure space $(\mu \geq 0)$. We say that a function $g: e \to \mathbf{R}$ is quasi-integrable iff it is μ-measurable and either g^+ or g^- is μ-integrable. Let f be a positive μ- measurable and non-negligible function (that is, we do not have $f = 0$ a.e.). Then

(i) if, for some $r > 0$, f^r is integrable, then $\log f$ is quasi-integrable.

(ii) let $r_0 > 0$ and suppose f^r is integrable for all $0 < r < r_0$. Let $A = \{x \in E: |f(x)| > 0\}$, and $N_r(f) = (\int |f|^r d\mu)^{1/r}$. Show that if $\mu(A) > 1$ then $\lim_{r \to 0} N_r(f) = +\infty$, and if $\mu(A) < 1$, then $\lim_{r\to 0} N_r(f) = 0$.

(iii) Suppose $\mu(A) = 1$. Show that

$$\lim_{r \to 0} \int |f|^r d\mu = 1$$

and $h(r) = \int |f|^r d\mu$ has a right hand derivative at $r = 0$ equal to $\int \log|f| d\mu$. Deduce that

$$\text{limit } N_r(f) = \exp(\int \log|f| d\mu).$$

11.5 <u>A capacity associated with a Radon Measure</u>

Let E be a locally compact σ-compact Hausdorff space and $\mu \in \mathcal{M}^+(E)$. Let

$$\mu^*(A) = \inf\{\mu(B): B \text{ is a } \mathcal{K}_\sigma \text{ and } B \supset A\}$$

for any $A \subset E$. Show that μ^* is a capacity on E (see §9 for the definition of capacity).

§12 SPACES OF MEASURES

We continue our study of Radon measures but now change the perspective. Namely we study $\mathcal{M}(E)$ as a whole, and investigate its topological and analytical structure. The main results are these: convergence properties in 12.2, compactness properties in 12.6, metrization in 12.10 and the theorem of approximation in 12.11. The last result makes use of the bipolar theorem, whose proof is postponed until §22. Some further results for $\mathcal{K}(E,\mathbb{R})$ and $\mathcal{M}(E)$ can be found in §16.

We begin with a topology for $\mathcal{M}(E)$.

12.1 <u>DEFINITION</u>. Let E be a locally compact Hausdorff space and $\mathcal{M}(E)$ the collection of Radon measures on E . For each $\varphi \in \mathcal{K}(E,\mathbb{R})$ consider the map $f_\varphi\colon \mathcal{M}(E) \to \mathbb{R}$ defined by $f(\mu) = \mu(\varphi)$. The initial topology on $\mathcal{M}(E)$ with respect to the maps $\{f_\varphi\colon \varphi \in \mathcal{K}(E,\mathbb{R})\}$ is called the <u>vague topology</u> (sometimes called the <u>weak*-topology</u>).

This topology is characterized by: for any net μ_i , $(\mu_i \to \mu)$ iff (for all $\varphi \in \mathcal{K}(E,\mathbb{R})$, $\mu_i(\varphi) \to \mu(\varphi))$. Thus the vague topology is also called the topology of <u>simple convergence</u>.

This topology arises from the following family of semi-norms: for each $\varphi \in \mathcal{K}(E,\mathbf{R})$, define

$$p_\varphi(\mu) = |\mu(\varphi)| .$$

Thus $\mathcal{M}(E)$ and $\mathcal{M}^+(E)$ have natural uniform structures derived from these semi-norms p_φ. As we saw in §11, $\mathcal{M}(E) = \mathcal{K}(E,\mathbf{R})'$ and if E is compact, $\mathcal{M}(E)$ is the dual of a Banach space. However on the dual we put the weak topology and not the norm topology.

Some basic properties are:

12.2 <u>THEOREM</u>. <u>Let</u> E <u>be a locally compact Hausdorff space</u>. <u>Then we have</u>:

(i) $\mathcal{M}^+(E)$ <u>is complete</u>

(ii) <u>suppose</u> $\mathcal{A} \subset \mathcal{K}(E,\mathbf{R})$ <u>is a linear subspace and for each compact</u> $K \subset E$, $\mathcal{A}^+ \cap \mathcal{K}_k(E,\mathbf{R})$ <u>is dense in</u> $\mathcal{K}_k^+(E,\mathbf{R})$. <u>Then</u>

(a) <u>any positive linear form</u> $\mu: \mathcal{A} \to \mathbf{R}$ <u>has a unique extension to a positive Radon measure</u>

(b) <u>if</u> $\mu_i \geq 0$ <u>is a net in</u> $\mathcal{M}^+(E)$ <u>and</u> $\mu_i(f)$ <u>converges for each</u> $f \in \mathcal{A}$, <u>then</u> μ_i <u>converges in</u> $\mathcal{M}^+(E)$

(c) <u>if</u> μ_i <u>is a net in</u> $\mathcal{M}(E)$ <u>and for each</u> $f \in \mathcal{K}(E,\mathbf{R})$, $\{|\mu_i(f)|\}$ <u>is bounded and</u> $\mu_i(\varphi)$ <u>converges for every</u> $\varphi \in \mathcal{A}$, <u>then</u> μ_i <u>converges in</u> $\mathcal{M}(E)$.

<u>Proof</u>. If μ_i is a net in $\mathcal{M}(E)$, then μ_i is Cauchy iff $\mu_i(f)$ is Cauchy for each $f \in \mathcal{K}(E,\mathbf{R})$. If $\mu_i \geq 0$ and $\mu_i(f)$ converges to say $\mu(f)$, then $\mu \in \mathcal{K}(E,\mathbf{R})^{*+} = \mathcal{M}^+(E)$. This proves (i). (Note that positivity is used crucially to conclude continuity of the limit by 11.2(ii).)

(ii)(a) Define a map $\mu^*: \mathcal{K}^+ \to \mathbf{R}$ by

$$\mu^*(f) = \sup\{\mu(g): g \in \mathcal{A}^+ \text{ and } 0 \leq g \leq f\}.$$

First, $\mu^*(f)$ is finite because for any $f \in \mathcal{K}^+$ there is a $g_0 \in \mathcal{A}^+$ such that $f \leq g_0$, (by the density assumption) and thus by positivity, $0 \leq \mu^*(f) \leq \mu(g_0)$.

To prove the assertion it suffices to show that μ^* is linear on $\mathcal{K}^+$, for it can then be extended to all of $\mathcal{K}$ by linearity.

Now we clearly have $\mu^*(\lambda f) = \lambda\mu^*(f)$. Next, let $f_1, f_2 \in \mathcal{K}^+$ and observe that

$$\mu^*(f_1 + f_2) = \sup\{\mu(g): g \in \mathcal{A}^+ \text{ and } 0 \leq g \leq f_1 + f_2\}$$
$$\geq \sup\{\mu(g_1 + g_2): g_1, g_2 \in \mathcal{A}^+,\ 0 \leq g_1 \leq f_1 \text{ and } 0 \leq g_2 \leq f_2\}$$
$$= \mu^*(f_1) + \mu^*(f_2)\ .$$

For the reverse inequality, note that for any compact K there is an M_K so that $S(g) \subset K$, $g \in \mathcal{A}^+$ implies $\mu(g) \leq M_K\|g\|$. Indeed find $g_0 \in \mathcal{A}^+$ so that $g_0 \geq 1$ on K and then $\mu(g) \leq \mu(g_0\|g\|) \leq \mu(g_0)\|g\|$, so we take $M_K = \mu(g_0)$.

Now given $\varepsilon > 0$, choose $g \in \mathcal{A}^+$ so that $g \leq f_1 + f_2$ and $\mu^*(f_1 + f_2) - \varepsilon \leq \mu(g)$. Since $\mathcal{K}$ is a lattice, $g = g_1 + g_2$, where $g_1, g_2 \in \mathcal{K}^+$ and $0 \leq g_1 \leq f_1$ and $0 \leq g_2 \leq f_2$. Now there exist g_1', $g_2' \in \mathcal{A}^+$ such that $\|g_1 - g_1'\| < \varepsilon$, $\|g_2 - g_2'\| < \varepsilon$ and $0 \leq g_1' \leq f_1$ and $0 \leq g_2' \leq f_2$ (see exercise 12.5).

Then we have $\mu(g) - \mu(g_1') - \mu(g_2') \leq 2M_K\,\varepsilon$ and so
$$\mu^*(f_1 + f_2) \leq \varepsilon + 2M_K\,\varepsilon + \mu(g_1') + \mu(g_2')$$
$$\leq \varepsilon + 2M_K\,\varepsilon + \mu^*(f_1) + \mu^*(f_2')\ .$$

Since $\varepsilon > 0$ is arbitrary, we have $\mu^*(f_1 + f_2) \leq \mu^*(f_1) + \mu^*(f_2)$ which completes the proof of (ii)(a).

(<u>Remark</u>: In §34, this same result and much more will be a special case of the theory of adapted spaces; see problem 34.2. For an easier, but important variant of this result see exercise 12.6).

For (b), given $K \subset E$ compact, there exists $f \in \mathcal{A}^+$ such that $f \geq 1$ on K (choose $\varphi \in \mathcal{K}^+(E,\mathbf{R})$ such that $\varphi = 2$ on K and approximate φ uniformly on K by some $f \in \mathcal{A}^+$). Thus, for i sufficiently large $\mu_i(f)$ is bounded and hence $\mu_i(1_K) \leq \mu_i(f)$ is bounded. Let $\varphi \in \mathcal{K}(E,\mathbf{R})$ and choose $\psi \in \mathcal{A}$ so that $\|\varphi - \psi\| < \varepsilon$. Then for i sufficiently large, if $S(\varphi) \cup S(\psi) \subset K$,

$$\begin{aligned}|\mu_i(\varphi) - \mu_j(\varphi)| &\leq |\mu_i(\varphi) - \mu_i(\psi)| \\ &\quad + |\mu_i(\psi) - \mu_j(\psi)| + |\mu_j(\psi) - \mu_j(\varphi)| \\ &\leq 2\, M_K\, \varepsilon + |\mu_i(\psi) - \mu_j(\psi)|\end{aligned}$$

where M_K is a bound for $\mu_i(1_K)$ for i sufficiently large. This estimate shows that $\mu_i(\varphi)$ converges and proves the assertion.

The proof of (c) proceeds the same way if there is a constant M_K such that $S(\varphi) \subset K$ implies $|\mu_i(\varphi)| \leq M_K\|\varphi\|$ independent of i. This is true according to the following:

12.3 THEOREM. Let $X \subset \mathcal{M}(E)$ be vaguely bounded that is, for each $f \in \mathcal{K}(E,\mathbf{R})$, $\{\mu(f): \mu \in X\}$ is a bounded set in $\mathbf{R}$. Then X is strongly bounded; that is, for each compact $K \subset E$ there is a constant M_K such that $S(f) \subset K$ implies $|\mu(f)| \leq M_K\|f\|$ for all $\mu \in X$.

Proof. Fix a compact $K \subset E$ and consider the Banach space $\mathcal{K}_K(E,\mathbf{R}) = \{f \in \mathcal{K}(E,\mathbf{R}): S(f) \subset K\}$. Consider the restriction of X to $\mathcal{K}_K$ as a family of continuous linear maps. By hypothesis, $\{|\mu(f)|: \mu \in X\}$ is bounded for each $f \in \mathcal{K}_K$. Consequently, by the uniform boundedness theorem, 7.4, the norms $\|\mu\|$ are uniformly bounded as elements of $\mathcal{K}_K'$, which proves the assertion. □

A theorem similar to 12.3 also holds for $\mathcal{K}^k(E,\mathbf{R})$ with $0 \leq k \leq \infty$ ($\mathcal{D} = \mathcal{K}^\infty$ is the Schwartz space) with a similar proof. (One replaces $\|f\|$ by $\|f\|_\ell$ the norm for f and its first ℓ derivatives, $\ell < \infty$.)

In general $\mathcal{M}(E)$ is not complete. In fact it is dense in $\mathcal{K}(E,\mathbf{R})^*$ (by 12.11 below) but does not equal it (Problem 12.4). However $\mathcal{M}(E)$ is sequentially complete:

12.4 COROLLARY. Let E be a locally compact Hausdorff space. Then $\mathcal{M}(E)$ is sequentially complete. That is, every Cauchy sequence in $\mathcal{M}(E)$ converges.

Proof. Let μ_n be a Cauchy sequence. Then for each $f \in \mathcal{K}(E,\mathbb{R})$, $\mu_n(f)$ is Cauchy and hence converges. Since μ_n is a sequence, $\{\mu_n\}$ is weakly bounded. Therefore, by 12.3, μ_n converges in $\mathcal{M}(E)$. $\square$

Notice that $\mu_i \to \mu$ does not imply that $\mu_i^+ \to \mu^+$. For example, if $x_n \to x$, $x_n \neq x$, let $\mu_n = \varepsilon_{x_n} - \varepsilon_x$ ($\varepsilon_x \in \mathcal{M}^+(E)$ is defined by $\varepsilon_x(f)$ $\varepsilon_x(f) = f(x)$; see 12.8). Then $\mu_n \to 0$ but $\mu_n^+ \to \varepsilon_x$.

Also without the boundedness assumption in 12.2 (ii-c), the theorem can fail. For example on $E = \mathbb{R}$, let $\mu_n = 2(n\,\varepsilon_{1/n} - n\,\varepsilon_{-1/n})$. Then for the continuously differentiable functions, $\mu_n(f)$ converges (to $f'(0)$), but μ_n does not converge to a measure. In this case the limit is a "distribution".

12.5 EXAMPLE (SCHWARTZ DISTRIBUTIONS). Let $\mathcal{K}^\infty$ denote the smooth functions with compact support in $E = \mathbb{R}^n$ (or on a smooth manifold). The sequential closure of $\mathcal{M}(E)$ in $\mathcal{K}^\infty(E,\mathbb{R})^*$ with the weak* topology (the initial topology for the maps $\mu \mapsto \mu(f)$) is called the space of distributions on E. Theorem 12.2 (ii) shows that positive distributions are in fact Radon measures. It also gives useful information on when a net of Radon measures converges to a Radon measure. Since $\mathcal{K}^\infty$ is not a lattice, every distribution is not the difference of two positive ones. It can be shown that if $\mathcal{K}^\infty$ is given the inductive limit topology (see §16) the distributions coincide with $(\mathcal{K}^\infty)'$ (an easy result using convolutions). More important, $(\mathcal{K}^\infty)'$ is sequentially complete (but not complete) by essentially the same proof as 12.4. (This is a basic theorem of L. Schwartz).

Remark. Let E be a locally compact Hausdorff space and $\mu \in \mathcal{M}(E)$. Then $\| |\mu| \| = \|\mu\|$ since $\|\mu\| = |\mu|(1)$ as we saw in 11.21.

The next theorem gives basic information about compact subsets of $\mathcal{M}(E)$ and is very useful in proving convergence theorems.

12.6 THEOREM. *Let* E *be a locally compact Hausdorff space, and* $X \subset \mathcal{M}(E)$. *Then the following are equivalent*:

(i) $c\ell(X)$ *is compact* (*in the vague topology*)

(ii) X *is vaguely bounded* (see 12.3)

(iii) *for each* $K \subset E$ *compact, there is an* N_K *such that for all* f *with support in* K

$$|\mu(f)| \leq N_K \|f\|$$

for all $\mu \in X$ (*that is,* X *is strongly bounded*.)

(iv) *for each* $K \subset E$ *compact there is a* P_K *such that*

$$|\mu|(K) \leq P_K$$

for all $\mu \in X$.

Proof. That (ii) is equivalent to (iii) was proven in 12.3. Now (iii) implies (iv) since we can find a compact neighborhood K' of K and $f \in \mathcal{K}(E,\mathbb{R})^+$ with $S(f) \subset K'$ and $f \geq 1_K$. Then since $\|\mu\| = \| \, |\mu| \, \| \leq N_{K'}$,

$$|\mu|(1_K) \leq |\mu|(f) \leq N_{K'}\|f\| = P_K$$

for all $\mu \in X$. Conversely,

$$|\mu(f)| \le |\mu|(|f|) \le |\mu|(\|f\| 1_K) = N_K \|f\|$$

where $N_K = |\mu|(K)$, by 11.21. So (iv) implies (iii).

Next we show that (iii) implies (i). In fact, let $\mathfrak{F}$ be an ultrafilter on $c\ell(X)$. Now $c\ell(X)$ is (vaguely) bounded so that for $f \in \mathcal{K}(E,\mathbb{R})$, the image of $\mathfrak{F}$ under the projection (evaluation) $\mu \mapsto \mu(f)$ is an ultrafilter on a bounded set in $\mathbb{R}$. Hence it converges to, say, $\mu_0(f)$. We easily see that μ_0 is a linear form on $\mathcal{K}(E,\mathbb{R})$, and by the estimate in (iii), $\mu_0 \in \mathcal{M}(E)$, and $\mathfrak{F}$ converges to μ_0 .

It is clear that (i) implies (ii) since the image of $c\ell(X)$ under each projection (evaluation; $\mu \mapsto \mu(\varphi)$) is compact and hence bounded. □

12.7 <u>COROLLARY</u>. <u>For</u> $a > 0$, $(a < \infty)$, <u>the set</u>

$$\{\mu \in \mathcal{M}(E) : \|\mu\| \le a\}$$

<u>is compact in</u> $\mathcal{M}(E)$. <u>If</u> E <u>is compact, then</u> $\{\mu \in \mathcal{M}^+(E) : \|\mu\| = a\}$ <u>is compact</u>.

<u>Proof</u>. First the set is closed, for if $\mu_i \to \mu$ and $0 \le f \le 1$, $f \in \mathcal{K}(E,\mathbb{R})$, then $|\mu_i|(f) \le a$

and hence $|\mu|(f) \leq a$. Therefore the norm of $|\mu|$, which equals $\|\mu\|$, is $\leq a$. Secondly, 12.5 (iv) shows that this set is compact. For the last statement observe that $\{\mu \geq 0: \|\mu\| = a\}$ is the inverse image in $\mathcal{M}^+$ of a by the continuous map $\mu \mapsto \mu(1)$. □

In general $\{\mu \in \mathcal{M}(E): \|\mu\| = a\}$ is not closed, although it is relatively compact. For example on $[-1,1]$, $\mu_n = \varepsilon_{1/n} - \varepsilon_{-1/n} \to 0$ but $\|\mu_n\| = 2$. (See 12.9 below for the definition of ε_x).

Another example of a compact set using 12.6 (iii) is (the proof being similar to 12.7):

12.8 <u>COROLLARY</u>. <u>For each</u> $\mu_o \in \mathcal{M}^+(E)$, <u>the set</u>

$$\{\mu \in \mathcal{M}(E): |\mu| \leq \mu_o\}$$ <u>is compact in</u> $\mathcal{M}(E)$.

12.9 <u>PROPOSITION</u>. <u>Let</u> E <u>be a locally compact Hausdorff space and</u> $x \in E$. <u>Let</u> $\varepsilon_x \in \mathcal{M}^+(E)$ <u>be given by</u> $\varepsilon_x(f) = f(x)$, called the <u>Dirac measure</u>. <u>The map</u> $x \mapsto \varepsilon_x$ <u>is a homeomorphism of</u> E <u>into</u> (<u>a subset of</u>) $\mathcal{M}^+(E)$ ($\mathcal{M}^+(E)$ <u>with the vague</u>

topology). Also if E is not compact, $\varepsilon_x \to 0$ as $x \to \infty$. (That is, for any $\varepsilon > 0$, $f \in \mathcal{K}(E,\mathbb{R})$ there is a compact set K so that $|\varepsilon_x(f)| < \varepsilon$ if $x \notin K$).

Proof. Clearly the map $x \mapsto \varepsilon_x$ is continuous ($x_i \to x$ implies $\varepsilon_{x_i} \to \varepsilon_x$) and is one to one, since E is locally compact Hausdorff, and hence completely regular. It is also easy to see that if $x_i \not\to x$ then $\varepsilon_{x_i} \not\to \varepsilon_x$ so that the map is a homeomorphism. The following argument can also be used: if E is compact then $x \mapsto \varepsilon_x$ is a homeomorphism (see §2). If E is non compact then let $E \cup \{\infty\} = c\ell(E)$ be its one point compactification. Clearly $x \to \infty$ implies $\varepsilon_x \to 0$, so the map $x \mapsto \varepsilon_x$ extends to a continuous map of $c\ell(E)$ into $\mathcal{M}^+(E)$, and so is a homeomorphism as $c\ell(E)$ is compact. □

An application of this is the following:

12.10 THEOREM. Let E be a locally compact Hausdorff space. Then $\mathcal{M}^+(E)$ with the vague topology, is metrizable and separable iff E has a countable base.

Remark. It is the topology and not the uniform structure on $\mathcal{M}^+(E)$ which is metrizable.

Proof. If $\mathcal{M}^+(E)$ is metrizable and separable, then so is E as it is homeomorphic to a subspace, by 12.8. Conversely, if E is second countable, it is σ-compact and we can write $E = \bigcup\{K_n : n = 1,2,\ldots\}$ where K_n are compact and $K_n \subset \operatorname{int}(K_{n+1})$. For each K_n, let $f_{n,k}$, $k = 1,2,\ldots$ be a family of functions dense in $\mathcal{K}^+(K_n,\mathbf{R})$, with $f_{n,k} \in \mathcal{K}(E,\mathbf{R})$. Then the vague topology on $\mathcal{M}^+(E)$ is the initial topology with respect to the maps:

$$\varphi_{n,k} : \mu \mapsto \mu(f_{n,k}) \in \mathbf{R} .$$

To see this we must show that if $\mu_i(f_{n,k}) \to \mu(f_{n,k})$ then $\mu_i(f) \to \mu(f)$ for all $f \in \mathcal{K}(E,\mathbf{R})$. But this follows from 12.2 (ii). Hence $\mathcal{M}^+(E)$ is homeomorphic to a subspace of $\mathbf{R}^{\mathbf{N}\times\mathbf{N}}$ which is metrizable and separable. □

Note that this argument fails for $\mathcal{M}(E)$. In fact $\mathcal{M}(E)$ is not in general metrizable (see §16).

A discrete measure is a measure of the form

$\mu = \sum_{i=1}^{n} a_i \, \varepsilon_{x_i}$. The support of μ is $\bigcup\{x_i\}$ and $\mu \geq 0$ iff $a_i \geq 0$ for all i .

12.11 THEOREM OF APPROXIMATION. Let E be a locally compact Hausdorff space. Then we have:

(i) the vector space generated by $\{\varepsilon_x\}$ for all $x \in E$ is dense in $\mathcal{M}(E)$,

(ii) the cone generated by the set of all $\{\varepsilon_x\}$ (that is, all sums $\sum_{i=1}^{n} a_i \, \varepsilon_{x_i}$ with $a_i \geq 0$) is dense in $\mathcal{M}^+(E)$,

(iii) for each compact K and $\mu \in \mathcal{M}^+(E)$ with $S_\mu \subset K$, there exist discrete μ_i with support a finite set in K , $\mu_i \geq 0$, $\|\mu_i\| = \|\mu\|$ and $\mu_i \to \mu$.

Proof. Although (i) follows from (ii), it can also be proved directly by a general density lemma in dual spaces; see Kelley [1] General Topology, page 108, ex. W(b).

To prove (ii) and (iii) we shall use another result whose proof is postponed until later. Let $B \subset \mathcal{M}(E)$ and define its polar $B^\circ \subset \mathcal{K}(E,\mathbf{R})$ by

$$B^\circ = \{f \in \mathcal{K}(E,\mathbf{R}) : \beta(f) \geq -1 \ \text{ for all } \ \beta \in B\}.$$

Also define, for $D \subset \mathcal{K}(E,\mathbf{R})$, its polar $D^\circ \subset \mathcal{M}(E)$ by

$$D^\circ = \{\alpha \in \mathcal{M}(E) : \alpha(f) \geq -1 \text{ for all } f \in D\}.$$

Then $B^{\circ\circ}$ is the <u>bipolar</u> of B and $B^{\circ\circ}$ is the closed convex hull of $B \cup \{0\}$ (intersection of all closed convex sets containing $B \cup \{0\}$). (This is not obvious; the proof will be given later in §22).

For (ii), let B be the cone in $\mathcal{M}^+(E)$; $X = \mathcal{K}(E,\mathbf{R})$ and note that

$$B^\circ = \{f \in \mathcal{K}(E,\mathbf{R}) : f \geq 0\}$$

and

$$B^{\circ\circ} = \{\mu \in \mathcal{K}(E,\mathbf{R})^* : \mu \geq 0\} = \mathcal{M}^+(E).$$

This proves (ii).

For (iii) we consider the restriction of μ to K. Since each $f \in \mathcal{K}(K,\mathbf{R})$ has an extension to $f \in \mathcal{K}(E,\mathbf{R})$ and $\mu(f)$ is independent of the extension (see 11.21), $\mu|K$ is well defined. Hence there is a net $\mu_i' \to \mu$ with $\mu_i' = \sum_{k=1}^{n_i} a_k \varepsilon_{x_k}$. Now we can replace μ_i' by $\mu_i = \|\mu\|\mu_i' / \|\mu_i'\|$ which converges to μ since $\|\mu_i'\| \to \|\mu\| < \infty$ (as K is compact we may apply μ to the function 1). □

Notice that $\mu_i \to \mu$ vaguely does not imply that $\|\mu_i - \mu\| \to 0$, or that $\|\mu_i\| \to \|\mu\|$. For example if $x_n \to x$ then $\varepsilon_{x_n} \to \varepsilon_x$ but if $x \neq x_n$, $\|\varepsilon_x - \varepsilon_{x_n}\| = 2$.

12.12 COROLLARY. *Let E be a locally compact Hausdorff space and $\mu \in \mathcal{M}(E)$ with $S(\mu)$ compact. Then there exists a net μ_i of discrete measures with $S(\mu_i) \subset S(\mu)$, $\|\mu_i\| = \|\mu\|$ and $\mu_i \to \mu$.*

Proof. Apply 12.10 (iii) to μ^+ and μ^-. □

PROBLEMS FOR §12.

12.1 Supports and convergence

Let E be a locally compact Hausdorff space.

(i) If $\mu \in \mathcal{M}(E)$ has finite support then $\mu = \Sigma\, a_i\, \varepsilon_{x_i}$ for some $a_i \in \mathbf{R}$, $x_i \in E$; $i = 1,\dots,n$, that is, μ is discrete.

If $E = \mathbf{R}$ and $\mathcal{K}(E,\mathbf{R})$ is replaced by the continuously differentiable functions with compact support, show that this assertion fails.

(ii) If $\mu_i \to \mu$ and $S(\mu_i) \subset K$ for a closed set $\mathcal{K}$ and all i, then $S(\mu) \subset K$. What can be said about "supported by"?

(iii) If $\mu_i \geq 0$ then $\mu_i \to 0$ iff $\mu_i(K) \to 0$ for every compact $K \subset E$. Can this be generalized?

12.2 Mappings defined by kernels

Let E be a locally compact space and $G : E \times E \to [0, +\infty]$ a lower semi-continuous function. For any μ in $\mathcal{M}(E)$, let G_μ be defined by $G_\mu(x) = \int G(x,y)d\mu(y)$.

(i) Show that G is the upper envelope of a family of continuous mappings of $E \times E$ into $[0,\infty]$ with compact support.

(ii) If G is finite, continuous, and with compact support, then the map $(x,\mu) \mapsto G_\mu(x)$ of $E \times \mathcal{M}^+$ into $\mathbb{R}$ is continuous ($\mathcal{M}^+$ with the vague topology).

(iii) For arbitrary G lower semi-continuous, the map $(x,\mu) \mapsto G_\mu(x)$ is lower semi-continuous.

(iv) If μ_n is a sequence in $\mathcal{M}^+$, $\mu_n \to \mu$, then

$$\liminf_{n\to\infty} G_{\mu_n}(x) \geq G_\mu(x) .$$

(v) Let $\mu \in \mathcal{M}^+(E)$ with compact support,

and Δ the diagonal in $E \times E$. If the restriction of G to the complement of Δ is finite and continuous, the map $x \mapsto G_\mu(x)$ is continuous outside the support of μ.

(vi) Let $G(x,y) = G(y,x)$ and show that for $\mu, \nu \geq 0$,

$$\int G_\mu \, d\nu = \int G_\nu \, d\mu .$$

(vii) Suppose E is compact. Let $\nu_0 \in \mathcal{M}^+$ and suppose G_{ν_0} is finite and continuous. Show that for any $\mu \geq 0$, G_μ is ν_0-integrable and the mappings $\mu \mapsto (G_\mu) \nu_0 ; \mu \mapsto \int G_\mu \, d\nu_0$ are continuous.

How would one extend this when E is only locally compact?

12.3 Separate and joint continuity of the evaluation map.

Let E be a compact Hausdorff space and let $\varphi : \mathcal{M}^+(E) \times \mathcal{K}(E,\mathbf{R}) \to \mathbf{R}$ be defined by $\varphi(\mu,f) = \mu(f)$. Show that φ is continuous (jointly). If $\mathcal{M}^+$ is replaced by $\mathcal{M}$, or E is not compact, show that one can deduce only sequential continuity of φ, and not continuity.

12.4 A linear form which is not a measure

Show that in general, $\mathcal{K}(E,\mathbf{R})^* \neq \mathcal{K}(E,\mathbf{R})'$, using for example $E = [0,1]$.

12.5 A Density Lemma

Let E be locally compact and $\mathcal{A} \subset \mathcal{K}$ satisfy the conditions of 12.2(ii). Show that for any $h \in \mathcal{K}^+$ and $\varepsilon > 0$ there exists $g \in \mathcal{A}^+$ such that $0 \leq g \leq h$ and $\|h-g\| \leq \varepsilon$.

12.6 An Extension Theorem

Let X be a compact space and $\mathcal{A} \subset \mathcal{C}(X,\mathbf{R})$ a dense sub-lattice vector space. Every positive linear form on $\mathcal{A}$ extends uniquely to a Radon measure on X. What if X is only locally compact?

12.7 Radon measures with compact support

Let E be a locally compact Hausdorff space.
(i) Show that $(\mathcal{C}(E,\mathbf{R})^*)^+$ can be identified with the positive Radon measures with compact support.
(ii) Let $\mathcal{C}_b(E,\mathbf{R})$ denote the bounded continuous functions. Show that $(\mathcal{C}_b(E,\mathbf{R})^*)^+$ is identifiable with the positive Radon measures of finite mass.

12.8 Comparable Radon Measures

(i) Let E be a locally compact Hausdorff space and $\mu, \nu \in \mathcal{M}^+(E)$. If $\mu \leq \nu$, $\|\nu\| = \|\mu\| < \infty$ then $\mu = \nu$.

(ii) If μ, ν are disjoint, then $\|\mu + \nu\| = \|\mu\| + \|\nu\|$.

§13 OPERATIONS ON MEASURES (I).

In this section we begin the study of some basic operations with measures. First we consider the product of a Radon measure with a function. Second we deal with the image of a Radon measure under a mapping, the key notion being that of a proper mapping. Finally we consider tensor products and deduce (easily) Fubini's theorem. In this connection, the theorem of approximation (12.10) is useful.

PRODUCT BY A FUNCTION

13.1 <u>DEFINITION</u>. Let E be a locally compact Hausdorff space, $\mu \in \mathcal{M}(E)$ and $f \in \mathcal{C}(E,\mathbf{R})$. Then define

$$\nu = f\mu : \mathcal{K}(E,\mathbf{R}) \to \mathbf{R} \text{ , by } \nu(\varphi) = \mu(f\varphi) .$$

More generally, let $\mu \in \mathcal{M}(E)$, and let $f : E \to [-\infty,\infty]$, be <u>locally</u> $|\mu|$-<u>integrable</u> (that is, for all $\varphi \in \mathcal{K}(E,\mathbf{R})$, $f\varphi$ is $|\mu|$-integrable). Then again define

$$\nu = f\mu : \mathcal{K}(E,\mathbf{R}) \to \mathbf{R} \text{ by } \nu(\varphi) = \mu(f\varphi),$$

and the restriction of ν to A by $\nu|A = 1_A \nu$.

13.2 PROPOSITION. (i) If $f : E \to [-\infty, \infty]$ is locally $|\mu|$-integrable then $\nu = f\mu \in \mathcal{M}(E)$, and (ii) for each $f \in \mathcal{C}(E,\mathbb{R})$, the map $\mu \mapsto f\mu$ of $\mathcal{M}(E)$ into $\mathcal{M}(E)$ is continuous.

Proof. (i) When $f \geq 0$, and $\mu \geq 0$, $\varphi \mapsto \nu(\varphi) = \mu(f\varphi)$ is a positive linear form (additive by 11.15) and so is a (Radon) measure. In the general case

$$f\mu = (f^+\mu^+ + f^-\mu^-) - (f^-\mu^+ + f^+\mu^-)$$

so again $f\mu$ is a measure.

(ii) The above shows that $f\mu \in \mathcal{M}(E)$. Now if $\mu_i \to \mu$ then for each $\varphi \in \mathcal{K}(E,\mathbb{R})$, $\mu_i(f\varphi) \to \mu(f\varphi)$ so that $f\mu_i \to f\mu$, or the map $\mu \mapsto f\mu$ is continuous. □

Notice that f is locally $|\mu|$-integrable iff f is $|\mu|$-integrable when restricted to every compact set iff f is $|\mu|$-integrable in some neighborhood of each point, and each implies that f is measurable.

Another useful remark is the following:

13.3 PROPOSITION. *Let* E *be a locally compact Hausdorff space and* $\mu \in \mathcal{M}(E)$. *Consider the set* $\{f \in \mathcal{K}(E,\mathbb{R}) : 0 \leq f \leq 1\}$ *as a net ordered by* $f \leq g$ *iff* $f(x) \leq g(x)$ *for all* $x \in E$. *Then* $f\mu$ *converges to* μ . *If* E *is second countable there is a countable subnet* f_n *with* $f_n\mu \to \mu$.

Proof. Let $\varphi \in \mathcal{K}(E,\mathbb{R})$ and choose $f_0 \in \mathcal{K}(E,\mathbb{R})$, $0 \leq f_0 \leq 1$ with $f_0 = 1$ on $S(\varphi)$. Then $f_0\mu(\varphi) = \mu(\varphi)$, and $f \geq f_0$ implies $f\mu(\varphi) = \mu(\varphi)$. Hence $f\mu \to \mu$.

For the second part, let $E = \bigcup\{F_n\}$ with F_n an increasing sequence of compact subsets, with $F_n \subset \mathrm{int}(F_{n+1})$. Choose f_n with $f_n = 1$ on F_n and $S(f_n) \subset F_{n+1}$. Then f_n is the required subnet. □

PROPER MAPPINGS. Recall that if E,F are Hausdorff topological spaces and $\Phi : E \to F$ is continuous, we say that Φ is a *proper* mapping iff for each $K \subset F$, compact, we have $\Phi^{-1}(K)$ is compact in E . Notice that a continuous map Φ is proper iff

$$\lim_{x\to\infty} \Phi(x) = \infty$$ where ∞ denotes the Alexandroff point at infinity (one point compactification).

[*Proof*. If Φ is proper and $K \subset F$ is compact, then $x \in E\setminus\Phi^{-1}(K)$ implies that $\Phi(x) \in E\setminus K$. Conversely, given any compact $K \subset F$ there is a compact $K' \subset E$ such that $x \in E\setminus K'$ implies $\Phi(x) \notin K$. Thus $\Phi^{-1}(K) \subset K'$, so $\Phi^{-1}(K)$ being closed, is compact.]

13.4 PROPOSITION. *Let E and F be locally compact Hausdorff spaces and* $\Phi : E \to F$ *a proper mapping. Then there is a unique continuous linear mapping*

$$\Phi^* : \mathcal{M}(E) \to \mathcal{M}(F)$$

such that for each $x \in E$,

$$\Phi^*(\varepsilon_x) = \varepsilon_{\Phi(x)} .$$

In fact, $\Phi^*(\mu) \cdot \varphi = \mu(\varphi \circ \Phi)$ *for each* $\varphi \in \mathcal{K}(F,\mathbf{R})$ *and* $\mu \in \mathcal{M}(E)$. *Also*, $\|\Phi^*(\mu)\| = \|\mu\|$, if $\mu \geq 0$, *and in general*, $\|\Phi^*(\mu)\| \leq \|\mu\|$.

Proof. Uniqueness is clear since the subspace spanned by $\{\varepsilon_x : x \in E\}$ is dense in $\mathcal{M}(E)$.

For existence, define $\Phi^*(\mu) \cdot \varphi = \mu(\varphi \circ \Phi)$ for $\mu \in \mathcal{M}(E)$ and $\varphi \in \mathcal{K}(E,\mathbf{R})$. Clearly $\Phi^*(\mu)$

is a linear map and $\mu \geq 0$ implies $\Phi^*(\mu) \geq 0$. Therefore for any $\mu \in \mathcal{M}(E)$, $\Phi^*(\mu) = \Phi^*(\mu^+) - \Phi^*(\mu^-) \in \mathcal{M}(F)$. The map $\Phi^*: \mathcal{M}(E) \to \mathcal{M}(F)$ is continuous, for $\mu_i \to \mu$ implies $\Phi^*(\mu_i)\cdot\varphi = \mu_i(\varphi \circ \Phi) \to \mu(\varphi \circ \Phi) = \Phi^*(\mu)\cdot\varphi$ for each $\varphi \in \mathcal{K}(F,\mathbf{R})$. From the definition of Φ^* it is clear that $\Phi^*(\varepsilon_x) = \varepsilon_{\Phi(x)}$.

For $\mu \in \mathcal{M}(E)$, we have

$$\begin{aligned}\|\Phi^*(\mu)\| &= \sup\{\Phi^*(\mu)\cdot\varphi : \|\varphi\| \leq 1\} \\ &= \sup\{\mu(\varphi \circ \Phi) : \|\varphi\| \leq 1\} \\ &\leq \sup\{\mu(f) : \|f\| \leq 1\} = \|\mu\| .\end{aligned}$$

If $\mu \geq 0$, then $\|\Phi^*(\mu)\| = \Phi^*(\mu)(1) = \sup\{\mu(\varphi \circ \Phi): \varphi \in \mathcal{K}(F,\mathbf{R}), 0 \leq \varphi \leq 1\}$. To prove that this equals $\mu(1)$, it suffices to show that for any $f \in \mathcal{K}(E,\mathbf{R})$, with $0 \leq f \leq 1$, there is a $\varphi \in \mathcal{K}(F,\mathbf{R})$. $0 \leq \varphi \leq 1$ such that $0 \leq f \leq \varphi \circ \Phi \leq 1$. However, if $K = \Phi(S(f))$, there is a $\varphi \in \mathcal{K}(F,\mathbf{R})$ such that $\varphi = 1$ on K and $0 \leq \varphi \leq 1$. This is the required φ and completes the proof. □

For μ not necessarily positive, $\|\Phi^*(\mu)\| \leq \|\mu\|$ generally cannot be replaced by an equality. For example on $\mathbf{R}$, let $\Phi: \mathbf{R} \to \mathbf{R}$ be defined by

$\Phi(x) = |x|$. Then if $\mu = \varepsilon_1 - \varepsilon_{-1}$, $\|\mu\| = 2$, but $\Phi^*(\mu) = 0$.

In practice we often want to take the image of μ under a non-proper map. In the case of abstract measures on σ-fields we do this as follows:

Let $\{E, \mathcal{A}, \mu\}$, $\mu \geq 0$ be a measure space ($\mathcal{A}$ is a σ-field on E and $\mu : \mathcal{A} \to [0,\infty]$ is a measure) and $\mathfrak{B}$ a σ-field on F . Let $\Phi: E \to F$ be measurable with respect to $\mathcal{A}$, $\mathfrak{B}$. Then define $\Phi^*(\mu)$, a measure on $\mathfrak{B}$ by

$$\Phi^*(\mu)(Y) = \mu(\Phi^{-1}(Y)) \leq \infty$$

for each $Y \in \mathfrak{B}$.

In the context of Radon measures, suppose E, F are locally compact spaces and $\Phi : E \to F$ is Borelian (for $U \subset F$ open, $\Phi^{-1}(U)$ is a Borel set in E). Let $\mu \in \mathcal{M}^+(E)$ and $\varphi \in \mathcal{K}^+(F,\mathbb{R})$. Then put

$$\Phi^*(\mu)\cdot\varphi = \mu(\varphi \circ \Phi) = \mu^*(\varphi \circ \Phi) .$$

If this is finite for each φ then clearly we have $\Phi^*(\mu) \in \mathcal{M}^+(F)$ ($\Phi^*\mu$ is additive since $\varphi \circ \Phi$ is measurable).

However, even if $\Phi^*(\mu)$ exists for a class

of μ (for instance $\{\mu \in \mathcal{M}^+(E): \|\mu\| < \infty\}$) , the map $\mu \mapsto \Phi^*(\mu)$ is not necessarily continuous.

For the next proposition we return to the setting of 13.4:

13.5 PROPOSITION. *Let* E, F, G *be locally compact Hausdorff spaces and* $\Phi: E \to F$ *and* $\Psi: F \to G$ *proper continuous mappings. Then*

(i) (*Functorality*) *for* $\mu \in \mathcal{M}(E)$,

$$(\Psi \circ \Phi)^*(\mu) = \Psi^*(\Phi^*(\mu)); \text{ that is,}$$
$$(\Psi \circ \Phi)^* = \Psi^* \circ \Phi^*$$

(ii) *if* Φ *is one to one then* $\Phi^*: \mathcal{M}(E) \to \mathcal{M}(F)$ *is also one to one*

(iii) *if* Φ *is onto* F *then* $\Phi^*(\mathcal{M}(E)) = \mathcal{M}(F)$; *that is,* Φ^* *is onto* $\mathcal{M}(F)$,

(iv) *for each* $\mu \in \mathcal{M}(E)$, $S(\Phi^*(\mu)) \subset \Phi(S(\mu))$ *with equality if* $\mu \geq 0$.

Proof. (i) is clear from the definition. To prove (ii) we first require a lemma:

13.6 LEMMA. *Let* E *and* F *be topological spaces with* F *locally compact and* $\Phi: E \to F$ *proper and continuous. Then* Φ *is closed; that is,*

($A \subset E$ closed) $\Rightarrow$ ($\Phi(A)$ closed).

Proof. Let $x_\alpha \to x$ where x_α is a net in $\Phi(A)$. Since F is locally compact there is a compact set K and α_0 so that $\alpha \geq \alpha_0$ implies $x_\alpha \in K$. Let $y_\alpha \in A \cap \Phi^{-1}(x_\alpha)$ so y_α is a net in the compact set $\Phi^{-1}(K)$. Thus there is a subnet converging to some $y \in A$. Since Φ is continuous, we have $\Phi(y) = x \in \Phi(A)$. □

Returning to the proof of 13.5 (ii), from the lemma we deduce that Φ is a homeomorphism onto its image and so Φ^* is a homeomorphism onto its image by (i). This proves (ii).

To prove (iii) it suffices to show that $\Phi^*(\mathcal{M}^+(E)) = \mathcal{M}^+(F)$. Let $\nu \in \mathcal{M}^+(F)$ and for $K \subset F$ compact, let ν_K be the restriction of ν to K. Find ν_K^i with finite support in K, positive and total mass $\|\nu_K\|$ and $\nu_K^i \to \nu_K$. Since Φ is onto, there is μ_K^i, a measure on $\Phi^{-1}(K)$ with total mass $\|\nu_K\|$, positive and $\Phi^*(\mu_K^i) = \nu_K^i$. Since $\{\mu_K^i\}$ form a bounded set (and hence lie in a compact), there is a convergent subnet, say to μ_K. By continuity, $\Phi^*(\mu_K) = \nu_K$.

Also, μ_K has total mass $\|\nu_K\|$ by 13.4. Now μ_K is a net for K increasing and $\{\mu_K : K \text{ compact}\}$ is bounded, for if $f \in \mathcal{K}^+(E,\mathbf{R})$, $0 \leq \mu_K(f) \leq \mu_K(1_{S(f)})\|f\| \leq \|\mu_K\|\cdot\|f\| \leq \|\nu\|\cdot\|f\|$ where $1_{S(f)}$ is the characteristic function of $S(f)$, the support of f. Hence μ_K has a convergent subnet, to say μ. Since $\nu_K \to \nu$ we have by continuity, $\Phi^*(\mu) = \nu$.

To prove (iv), we show at first that $F \setminus \Phi(S(\mu)) \subset F \setminus S(\Phi^*(\mu))$. Since $F \setminus \Phi(S(\mu))$ is open, we have that $f \in \mathcal{K}(F,\mathbf{R})$, $S(f) \subset F \setminus \Phi(S(\mu))$ implies $\Phi^*(\mu)(f) = 0$, that is $\mu(f \circ \Phi) = 0$, which proves our claim. That is, $S(\Phi^*(\mu)) \subset \Phi(S(\mu))$. Finally we want to show equality if $\mu \geq 0$. Note that if $f \in \mathcal{K}^+(E,\mathbf{R})$ and $f(y) > 0$ then $\mu(f) = 0$ implies $y \notin S(\mu)$. (<u>Proof</u>. There is a neighborhood U of y on which $f \geq \varepsilon$ for $\varepsilon > 0$. Then for any $g \geq 0$, $S(g) \subset U$, $g \leq Mf$ for a constant M. Thus $0 \leq \mu(g) \leq M\,\mu(f) = 0$, so $\mu(g) = 0$. Thus μ is zero on U, so U lies in the complement of $S(\mu)$.) Now $x \notin S(\Phi^*(\mu))$ implies there is an $f \in \mathcal{K}^+(F,\mathbf{R})$ with $f(x) > 0$ and supported in the complement of $S(\Phi^*(\mu))$ by Urysohn's lemma. Thus $\Phi^*(\mu)(f) = \mu(f \circ \Phi) = 0$. If $y \in \Phi^{-1}(x)$, then $f \circ \Phi(y) > 0$ and $\mu(f \circ \Phi) = 0$ which implies $y \notin S(\mu)$ by our remark above. This means that $x \notin \Phi(S(\mu))$ and hence $\Phi(S(\mu)) \subset S(\Phi^*(\mu))$ proving the assertion. □

<u>Remarks</u>. (i) the proof of (iii) also shows that if $A \subset \mathcal{M}^+(E)$ is closed, then $\Phi^*(A)$ is closed. Indeed, if $\nu_i = \Phi^*(\mu_i)$ and $\nu_i \to \nu$, $\nu_i \in A$, then the family $\{\mu_{iK}\}$ for K compact is bounded and has a cluster point, say $\mu \in A$. Then $\nu = \Phi(\mu)$.

(ii) For $\mu \in \mathcal{M}(E)$, one need not have equality in (iv). For example, let $E = F = \mathbb{R}$ and $\Phi(x) = |x|$, $\mu = \varepsilon_1 - \varepsilon_{-1}$. Then $\Phi(S(\mu)) = \{1\}$, but $\Phi^*(\mu) = 0$.

TENSOR PRODUCTS

We turn now to tensor products, the main theorem being the following.

13.7 <u>THEOREM</u>. <u>Let</u> E <u>and</u> F <u>be locally compact Hausdorff spaces</u>. <u>Then there is a unique mapping</u>

$$\varphi : \mathcal{M}(E) \times \mathcal{M}(F) \to \mathcal{M}(E \times F)$$

<u>denoted</u> $\varphi(\mu,\nu) = \mu \otimes \nu$ <u>such that</u>

(i) φ <u>is bilinear</u>

(ii) φ *is continuous in each variable separately* (*not jointly*)

(iii) *for each* $(x,y) \in E \times F$ *we have*

$$\varphi(\varepsilon_x, \varepsilon_y) = \varepsilon_{(x,y)}.$$

Furthermore, $\mu \otimes \nu$ *is the unique measure such that for any* $f \in \mathcal{K}(E,\mathbb{R})$, $g \in \mathcal{K}(F,\mathbb{R})$,

$$(\mu \otimes \nu)(f \otimes g) = \mu(f)\cdot\nu(g)$$

where $(f \otimes g)(x,y) = f(x)\cdot g(y)$.

Also, for $\Psi \in \mathcal{K}(E \times F,\mathbb{R})$ *we have*

$$\begin{aligned}(\mu \otimes \nu)(\Psi) &= \int\left(\int \Psi(x,y)\,d\mu(x)\right)d\nu(y) \\ &= \int\left(\int \Psi(x,y)\,d\nu(y)\right)d\mu(x) .\end{aligned}$$

For the proof we shall proceed in several steps:

13.8 *LEMMA*. *If* E *and* F *are compact Hausdorff spaces then the algebra* $\mathcal{K}(E,\mathbb{R}) \otimes \mathcal{K}(F,\mathbb{R})$ *is dense in* $\mathcal{K}(E \times F,\mathbb{R})$. ($\mathcal{K}(E,\mathbb{R}) \otimes \mathcal{K}(F,\mathbb{R})$ *consists of functions of the form* $\sum_{i=1}^{n} f_i(x)\cdot g_i(y)$) . *Furthermore, if* E *and* F *are locally compact Hausdorff spaces and* $\Psi \in \mathcal{K}(E \times F,\mathbb{R})$ *and* $\varepsilon > 0$, *there exist* $f_1,\ldots,f_n \in \mathcal{K}(E,\mathbb{R})$, $g_1,\ldots,g_n \in \mathcal{K}(F,\mathbb{R})$ *such* that $\|\Psi - \Sigma f_i \otimes g_i\| < \varepsilon$.

Proof. The first part is immediate by the Stone Weierstrass Theorem ($\mathcal{K}(E,\mathbf{R}) \otimes \mathcal{K}(F,\mathbf{R})$ separates points by Urysohn's Lemma). For the second part, let E_1, F_1 be compact with $S(\psi) \subset E_1 \times F_1$. Let E_2, F_2 be compact neighborhoods of E_1, F_1 respectively. Choose $f_1', \ldots, f_n' \in \mathcal{K}(E_2,\mathbf{R})$, $g_1', \ldots, g_n' \in \mathcal{K}(F_2,\mathbf{R})$ such that $\|\psi - \Sigma f_i' \otimes g_i'\| < \varepsilon$ considered as functions in $E_2 \times F_2$. Let $f_0 : E_2 \to [0,1]$ be 1 on E_1 and have support $S(f_0) \subset \operatorname{int}(E_2)$. Define $g_0 : F_2 \to [0,1]$ similarly. Then the functions $f_0 \cdot f_i'$ and $g_0 \cdot g_i'$ have obvious extensions to $f_i \in \mathcal{K}(E,\mathbf{R})$ and $g_i \in \mathcal{K}(F,\mathbf{R})$, respectively, so that $\|\psi - \Sigma f_i \otimes g_i\| < \varepsilon$. □

Remark. We next show that the various maps in 13.7 are unique as a prelude to showing they are identical.

13.9 LEMMA. (i) φ in 13.7 is unique if it exists.

(ii) there is at most one bilinear mapping $\Phi : \mathcal{M}(E) \times \mathcal{M}(F) \to \mathcal{M}(E \times F)$ such that for all $f \in \mathcal{K}(E,\mathbf{R})$, $g \in \mathcal{K}(F,\mathbf{R})$, we have

$$\Phi(\mu,\nu)(f \otimes g) = \mu(f)\nu(g).$$

<u>Proof</u>. Since φ is bilinear and continuous in each variable separately, and the space generated by $\{\varepsilon_x\}$ is dense, (i) easily follows.

To prove (ii), note that if $\psi \in \mathcal{K}(E \times F, \mathbf{R})$ and $\varepsilon > 0$, there are $f_k \in \mathcal{K}(E, \mathbf{R})$, $g_k \in \mathcal{K}(F, \mathbf{R})$ with

$$\|\psi - \sum_{k=1}^{n} f_k \otimes g_k\| < \varepsilon$$

and all supports in a fixed compact set H, by 13.8. Therefore since $\Phi(\mu,\nu) \in \mathcal{M}(E \times F)$ is assumed,

$$|\Phi(\mu,\nu)(\psi) - \Phi(\mu,\nu)(\sum_{k=1}^{n} f_k \otimes g_k)| < M_H \varepsilon$$

for a constant M_H. This proves (ii). □

13.10 <u>LEMMA</u>. <u>The (unique) mapping given in</u> 13.9 (ii) <u>exists and is given by</u>

$$\begin{aligned} \mu \otimes \nu(\Psi) &= \int (\int \psi(x,y) d\mu(x)) d\nu(y) \\ &= \int (\int \psi(x,y) d\nu(y)) d\mu(x) \end{aligned}$$

<u>for all</u> $\psi \in \mathcal{K}(E \times F, \mathbf{R})$.

<u>Proof</u>. Observe that the map $y \mapsto \int \psi(x,y)\, d\mu(x)$ is an element of $\mathcal{K}(F, \mathbf{R})$ by uniform continuity of ψ,

so that these formulas make sense. Clearly both are such that $(\mu \otimes \nu)(f \otimes g) = \mu(f)\cdot\nu(g)$, so by uniqueness, they are equal. (If $S(\psi) \subset K_1 \times K_2$ then $|\mu \otimes \nu(\psi)| \leq M_{K_1}^{(\mu)} M_{K_2}^{(\nu)} \|\psi\|$ in either case, so that these are Radon measures). □

<u>Proof of 13.7</u>. We define φ by the formulas in 13.10. Clearly φ is bilinear. From the first formula we have continuity in ν and from the second, continuity in μ . Also, $\varepsilon_x \otimes \varepsilon_y = \varepsilon_{(x,y)}$.

This, with 13.9 (i) proves the first part of 13.7. The second and third parts follow from 13.9 (ii) and 13.10. □

<u>Remarks</u>. (i) Notice that the above includes the proof of Fubini's theorem for $\psi \in \mathcal{K}(E \times F,\mathbb{R})$, which gives the general case for $\mu , \nu \geq 0$ (see problem 13.2).

(ii) Let E and F be compact and $\pi : E \times F \to E$ the projection. Then $\pi(\mu \otimes \nu)$ is not in general μ . In fact it is, from 13.7,

$$\pi(\mu \otimes \nu) = \nu(1)\mu .$$

In particular if $\nu(1) = 0$, $\pi(\mu \otimes \nu) = 0$.

Next we investigate two important cases in which $\otimes$ is jointly continuous. First we prepare:

13.11 LEMMA. *Let* E *be a locally compact Hausdorff space and* $\mu_i \in \mathcal{M}^+(E)$, $\mu_i \to \mu$, μ_i *a net. Then for all* $f \in \mathcal{K}(E,\mathbf{R})$, *there is an* i_0 *such that* $\{f\mu_i : i \geq i_0\}$ *is a bounded set in* $\mathcal{M}^+(E)$ (see 12.3).

Proof. We may suppose $f \geq 0$. Given $\varphi \in \mathcal{K}^+(E,\mathbf{R})$, $0 \leq \mu_i(f\varphi) \to \mu(f\varphi)$. Note that $\mu_i(f\varphi) \leq \mu_i(f\|\varphi\|) = \|\varphi\|\mu_i(f)$, which is bounded for i large enough, as $\mu_i(f) \to \mu(f)$. □

13.12 PROPOSITION. *Let* E *and* F *be locally compact Hausdorff spaces. Then*

(i) *if* $X \subset \mathcal{M}(E)$ *and* $Y \subset \mathcal{M}(F)$ *are bounded sets* (see 12.3) *then* $\otimes$ *restricted to* $X \times Y$ *is* (*jointly*) *continuous.*

(ii) $\otimes$ *restricted to* $\mathcal{M}^+(E) \times \mathcal{M}^+(F)$ *is* (*jointly*) *continuous.*

Proof. First we prove (i). Suppose $\mu_i \to \mu$ and $\nu_i \to \nu$. We must show that $\mu_i \otimes \nu_i \to \mu \otimes \nu$ vaguely. Let $f \in \mathcal{K}(E \times F,\mathbf{R})$ and $S(f) \subset \operatorname{int}(K) \times \operatorname{int}(K')$ for $K \subset E$, $K' \subset F$, compact.

For $\varepsilon > 0$ find $f_k \in \mathcal{K}(E,\mathbf{R})$ such that $S(f_k) \subset K$, $g_k \in \mathcal{K}(F,\mathbf{R})$ such that $S(g_k) \subset K'$; $k = 1,\ldots,n$ with

$$\|f - \sum_{k=1}^{n} f_k \otimes g_k\| < \varepsilon .$$

Choose i_o such that $i \geq i_o$ implies

$$\sum_{k=1}^{n} |\mu_i(f_k)|\ |\nu_i(g_k) - \nu(g_k)| < \varepsilon$$

and

$$\sum_{k=1}^{n} |\nu(g_k)|\ |\mu_i(f_k) - \mu(f_k)| < \varepsilon$$

for $k = 1,\ldots,n$. This is possible because $\Sigma|\mu_i(f_k)|$ is bounded for i large enough. Now for $h \in \mathcal{K}(E \times F,\mathbf{R})$, $S(h) \subset K \times K'$ implies

$$|\mu_i \otimes \nu_i(h)| \leq M_K^{(\mu)} M_{K'}^{(\nu)} \|h\| \leq M_{K,K'} \|h\|$$

where $M_{K,K'}$ is a constant independent of μ,ν since X and Y are bounded. Therefore, for $i \geq i_o$, we have the following estimate:

$$
\begin{aligned}
|\mu_i \otimes \nu_i(f) - \mu\otimes\nu(f)| &\leq |\mu_i \otimes\nu_i(f) - \Sigma\mu_i\otimes\nu_i(f_k\otimes g_k)| \\
&+ |\Sigma\mu_i\otimes\nu_i(f_k\otimes g_k) - \Sigma\mu_i\otimes\nu(f_k\otimes g_k)| \\
&+ |\Sigma\mu_i\otimes\nu(f_k\otimes g_k) - \Sigma\mu\otimes\nu(f_k\otimes g_k)| \\
&+ |\Sigma\mu\otimes\nu(f_k\otimes g_k) - \mu\otimes\nu(f)| \\
&\leq M_{K,K'}\varepsilon + \varepsilon + \varepsilon + M_{K,K'}\varepsilon = 2\varepsilon M_{K,K'} + 2\varepsilon .
\end{aligned}
$$

This proves the result (i). For (ii) we use the lemma 13.11 to reduce the procedure to that in (i). □

The question of the support and norm of $\mu\otimes\nu$ is answered as follows:

13.13 PROPOSITION. Let E and F be locally compact Hausdorff spaces, and $(\mu,\nu) \in \mathcal{M}(E)\times\mathcal{M}(F)$. Then we have:

(i) $S(\mu\otimes\nu) = S(\mu)\times S(\nu)$

(ii) if $\nu_1,\nu_2 \geq 0$ are disjoint and $\mu \geq 0$, then $\mu\otimes\nu_1$ and $\mu\otimes\nu_2$ are disjoint.

(iii) $|\mu\otimes\nu| = |\mu|\otimes|\nu|$

(iv) $\|\mu\otimes\nu\| = \|\mu\|\cdot\|\nu\|$.

Proof. (i) If $(x,y) \notin S(\mu\otimes\nu)$ then either $x \notin S(\mu)$ or $y \notin S(\nu)$. Indeed, if $x \in S(\mu)$, for any neighborhood U of x there exists

$f \in \mathcal{K}(E,\mathbf{R})$ such that $\mu(f) \neq 0$ and $S(f) \subset U$. Choose $(x,y) \in U \times V \subset E \times F \setminus S(\mu \otimes \nu)$. Then $g \in \mathcal{K}(F,\mathbf{R})$, $S(g) \subset V$ implies $\mu(f)\cdot\nu(g) = 0$, so $\nu(g) = 0$. Hence $y \notin S(\nu)$.

Conversely, $x \notin S(\mu)$, or $y \notin S(\nu)$ implies $(x,y) \notin S(\mu \otimes \nu)$. Indeed, if $x \notin S(\mu)$ there is a neighborhood U of x on which μ vanishes. Then for any $g \in \mathcal{K}(F,\mathbf{R})$, $S(f) \subset U$, $(\mu\otimes\nu)(f\otimes g) = 0$. Hence $\mu \otimes \nu$ vanishes on $U \times F$. This proves (i).

To prove (ii), we use the criterion for disjointness of 10.11. (One could also use the Jordan Decomposition theorem here but it is not necessary). Let $\psi \in \mathcal{K}(E \times F,\mathbf{R})^+$ and $\varepsilon > 0$. Choose $f_1,\ldots,f_n \in \mathcal{K}(E,\mathbf{R})^+$ and $g_1,\ldots,g_n \in \mathcal{K}(F,\mathbf{R})^+$ all with supports in fixed compacts K_1 and K_2 such that $\|\psi - \Sigma f_k g_k\| < \varepsilon$ and $S(\varphi) \subset K_1 \times K_2$ implies $|\mu \otimes \nu_2(\varphi)| \leq M\|\varphi\|$. Now write (by 10.11), $g_k = g_k^1 + g_k^2$ so that $\nu_1(g_k^1) + \nu_2(g_k^2) < \varepsilon/\Sigma\,\mu\,(f_k)$. Then $\psi = \Sigma f_k g_k^1 + \Sigma f_k g_k^2 + \varphi$ where $\|\varphi\| < \varepsilon$ so we obtain $\mu \otimes \nu_1(\Sigma f_k g_k^1) + \mu \otimes \nu_2(\Sigma f_k g_k^2 + \varphi)$

$$= \{\Sigma\mu(f_k)\}(\nu_1(g_k^1) + \nu_2(g_k^2)) + \mu \otimes \nu_2(\varphi) \leq \varepsilon + \varepsilon M$$

which proves the assertion.

For (iii), write $\mu = \mu^+ - \mu^-$ and $\nu = \nu^+ - \nu^-$

and note $\mu \otimes \nu = (\mu^+ \otimes \nu^+ + \mu^- \otimes \nu^-) - (\mu^+ \otimes \nu^- + \mu^- \otimes \nu^+)$. By (ii), $\mu^+ \otimes \nu^+$ and $\mu^+ \otimes \nu^-$ are disjoint. In fact, we conclude that $(\mu \otimes \nu)^+ = \mu^+ \otimes \nu^+ + \mu^- \otimes \nu^-$ and $(\mu \otimes \nu)^- = \mu^+ \otimes \nu^- + \mu^- \otimes \nu^+$. (See exercise 10.1 (ii)). Therefore

$$|\mu \otimes \nu| = (\mu \otimes \nu)^+ + (\mu \otimes \nu)^- = (\mu^+ + \mu^-) \otimes (\nu^+ + \nu^-)$$

as required.

For (iv) we have

$$\|\mu \otimes \nu\| = \|(\mu \otimes \nu)^+\| + \|(\mu \otimes \nu)^-\| = \|\mu\| \, \|\nu\| ,$$

by using the formula above for $(\mu \otimes \nu)^+$ applied to the function 1. □

As a corollary to this proof, we have

$$(\mu \otimes \nu)^+ = \mu^+ \otimes \nu^+ + \mu^- \otimes \nu^-$$

and

$$(\mu \otimes \nu)^- = \mu^+ \otimes \nu^- + \mu^- \otimes \nu^+ .$$

Further, both these sums are disjoint. That is, $\mu^+ \otimes \nu^+$ and $\mu^- \otimes \nu^-$ are disjoint. (By an argument like (ii) above).

For the product of a finite number of measures, we can define $\mu_1 \otimes \mu_2 \otimes \dots \otimes \mu_n$ by induction. For infinite products however, it is necessary to make some restrictions as follows.

13.14 <u>THEOREM</u>. <u>Let</u> $\{E_i\}_{i \in I}$ <u>be a family of compact Hausdorff spaces and</u> $\mu_i \in \mathcal{M}^+(E)$ <u>with</u> $\|\mu_i\| = 1$. <u>Then there is a unique measure</u> $\mu = \otimes \mu_i \in \mathcal{M}^+(\Pi E_i)$ <u>with</u> $\|\mu\| = 1$ <u>such that for any finite set of indices</u> J <u>where</u> $i \in J \subset I$, <u>with</u> $\pi_J : E = \prod_{i \in I} E_i \to \prod_{i \in J} E_i$ <u>the projection, we have</u> $\pi_J(\mu) = \mu_J = \bigotimes_{i \in J} \mu_i$.

<u>Proof</u>. For uniqueness, observe that if $f_i \in \mathcal{K}(E_i, \mathbf{R})$ and J is finite then $\mu(\bigotimes_{i \in J} f_i \bigotimes_{i \notin J} 1) = \prod_{i \in J} \mu_i(f_i)$, and functions of the form $\bigotimes_{i \in J} f_i$ are dense in $\mathcal{K}(\prod_{i \in J} E_i, \mathbf{R})$, by Stone-Weierstrass. Clearly if μ exists, then $\|\mu\| = 1$, (since $\|\mu\| = \mu(1)$) .

For existence, for each $J \subset I$, J finite, choose $x_{I \setminus J} \in \prod_{i \in I \setminus J} E_i$ and define

$$\mu_J = \bigotimes_{i \in J} \mu_i \otimes \varepsilon_{x_{I \setminus J}}$$

on $E = \prod_{i \in J} E_i \times \prod_{i \in I \setminus J} E_i$. Now for J ordered by inclusion, μ_J is a net on $\{\mu \in \mathcal{M}^+(E): \|\mu\| \leq 1\}$ which is compact. Hence some subnet converges, to μ , say. By Remark (ii) preceding 13.11, μ has the desired projection property. □

One can draw a similar conclusion if $\mu_i \geq 0$ and $\|\mu_i\| = 1 + \alpha_i$ with $\Pi(1 + \alpha_i)$ convergent by considering $\mu_i/\|\mu_i\|$. (Exercise for the reader.)

Remarks. (i) If μ_i are not positive on E_i the theorem fails because of the property $\pi(\mu \otimes \nu) = \nu(1)\mu$ mentioned earlier. For example on $[-1,1] = E_n, n \in \mathbb{N}$, let μ_n be +(Lebesgue measure) on $[0,1]$ and -(Lebesgue measure on) $[-1,0]$. Then $\bigotimes_{n=1}^{\infty} \mu_n$ constructed in the proof exists, but is identically zero!

(ii) If the E_i are only locally compact (for example $E_i = \mathbb{R}$, $i \in \mathbb{N}$) then ΠE_i need not be locally compact, so $\otimes \mu_i$ can't be a Radon measure. A product can be defined however in another setting (involving "cylinder measures"). See, for example, Halmos, [2] Measure Theory, §38.

PROBLEMS FOR §13

13.1 Equivalence of the definitions of tensor product

Show that, for $\mu, \nu \geq 0$, $\mu \otimes \nu$ as defined in 13.7 coincides with the product as Borel measures (see §3).

13.2 Fubini's Theorem

Deduce Fubini's theorem for integrable functions from 13.7.

13.3 Images of measures

Let E and F be locally compact spaces and $\Phi : E \to F$ be a proper mapping. If $f : F \to [0,\infty]$ is lower semi-continuous, show that $\Phi^*(\mu)\cdot f = \mu(f \circ \Phi)$. What can you say about a more general class of functions?

13.4 Cancellation law for tensor products

Let E and F be locally compact spaces and $\mu \in \mathcal{M}(E)$, $\nu \in \mathcal{M}(F)$. If $\mu \otimes \nu = 0$ then either $\mu = 0$ or $\nu = 0$.

13.5 Continuity of the inf mapping

(i) Let E be a locally compact Hausdorff

space. For $\mu_i, \nu_i \in \mathcal{M}(E)$, $\mu_i \to \mu$ and $\nu_i \to \nu$ show that $\mu_i \leq \nu_i$ implies $\mu \leq \nu$.

(ii) Now let E be compact. If $\mu_i \to \mu$ does $\mu_i \wedge \nu \to \mu \wedge \nu$? Show that if $\|\mu_i - \mu\| \to 0$ and $\|\nu_i - \nu\| \to 0$ then $\mu_i \wedge \nu_i \to \mu \wedge \nu$.

§14 OPERATIONS ON MEASURES (II)

In this section we conclude our study of the basic operations on measures. We consider convolutions of measures, the associated results about Haar measures, and finally a brief sketch of diffusions.

CONVOLUTIONS

To introduce the notion of convolution (and Fourier analysis) one requires a group; more specifically a <u>topological group</u>, that is, a group G which is also a Hausdorff topological space such that the map $f: G \times G \to G$ defined by $f(x,y) = x \cdot y^{-1}$ is continuous, $x \cdot y$ denoting group multiplication.

The basic fact is the following:

14.1 <u>THEOREM</u>. <u>Let</u> G <u>be a locally compact</u> (<u>topological</u>) <u>group</u>. <u>Then for any</u> $K \subset G$, K <u>compact</u>, <u>there is a unique mapping</u> $* : \mathcal{M}_K(G) \times \mathcal{M}(G) \to \mathcal{M}(G)$; $(\mu,\nu) \mapsto \mu * \nu$, (<u>where</u> $\mathcal{M}_K(G)$ <u>denotes</u> $\{\mu \in \mathcal{M}(G);\ S(\mu) \subset K\}$), <u>such that</u>

(i) $\mu * \nu$ <u>is continuous in each variable separately</u>

(ii) $\mu * \nu$ <u>is bilinear</u>

(iii) $\varepsilon_x * \varepsilon_y = \varepsilon_{xy}$.

<u>In fact,</u> $\mu * \nu = \varphi(\mu \otimes \nu)$, <u>where</u> $\varphi: G \times G \to G$ <u>is defined by</u> $\varphi(x,y) = xy$ <u>and we write</u> $\varphi^*(\pi) = \varphi(\pi)$. <u>We call</u> $\mu * \nu$ <u>the convolution of</u> μ <u>and</u> ν .

<u>Proof</u>. Uniqueness is clear as in 13.7. For existence, consider $\mu * \nu$ defined by $\varphi(\mu \otimes \nu)$. By 13.4, 13.7, we shall have the result if $\varphi|K \times G$ is proper. To see that this is the case, let $K' \subset G$ be compact and note that $\varphi^{-1}(K') \subset K \times K^{-1}K'$ where $K^{-1}K' = \{x^{-1}y : x \in K, y \in K'\}$, which is compact. □

If $S(\nu)$ is compact, we let $\mu * \nu = \varphi(\mu \otimes \nu)$ in a similar way.

For positive measures we can define $\mu * \nu = \varphi(\mu \otimes \nu)$ even if μ and ν don't have compact support, as long as $\varphi(\mu \otimes \nu)$ is a Radon measure (is finite when applied to elements of $\mathcal{K}(G,\mathbb{R})$). However, $\mu * \nu$ will not be (separately) continuous in this general case.

14.2 <u>THEOREM</u>. <u>Let</u> G <u>be a locally compact group,</u> <u>and</u> $K \subset G$ <u>a compact set</u>. <u>Then</u>

(i) $*$ _restricted to_ $\mathcal{M}_K^+(G) \times \mathcal{M}^+(G)$ _is jointly continuous_,

(ii) _if_ $X \subset \mathcal{M}_K(G)$ _and_ $Y \subset \mathcal{M}(G)$ _are bounded sets_ (_see_ 12.3) _then_ $*$ _restricted to_ $X \times Y$ _is jointly continuous_.

Proof. Clear from 13.4 and 13.12. □

Remark. These theorems can be generalized in a routine way to Schwartz distributions on Lie groups (a Lie group is a topological group which also has a differentiable structure). See 12.4.

14.3 _COROLLARY_. _Let_ G _be a locally compact group_, $\mu_n \to \mu$ _and_ $\nu_n \to \nu$ _be convergent sequences in_ $\mathcal{M}(G)$. _Suppose that the_ μ_n _all have supports in some compact set_. _Then_ $\mu_n * \nu_n \to \mu * \nu$.

Proof. Convergent sequences are bounded sets. □

Notice that in the corollary we are again using the fact that "weakly bounded" and "strongly bounded" are equivalent.

Another useful case when $*$ is jointly continuous is the following:

14.4 PROPOSITION. *Let* G *be a locally compact group and* $\mu_i \in \mathcal{M}^+(G)$ *a family of probability measures (that is,* $\|\mu_i\| = 1$) *with the property that: for all* $\varepsilon > 0$ *there is a compact set* $K \subset G$ *with* $\mu_i(G\setminus K) < \varepsilon$ *for all* i. *Let* $\nu_j \in \mathcal{M}^+(E)$ *with* $\|\nu_j\|$ *bounded uniformly in* j. *Then the map* $(\mu_i, \nu_j) \mapsto \mu_i * \nu_j$ *is jointly continuous.*

Proof. By 14.2, for each compact K, the map $(\mu_i|K, \nu_j) \mapsto (\mu_i|K) * \nu_j$ is continuous. Let $f \in \mathcal{K}(G)$ and $\varepsilon > 0$. Choose K as in 14.4 and if $\mu_i \to \mu$, $\nu_j \to \nu$, choose i_0, j_0 so that $i \geq i_0$, $j \geq j_0$ implies $|(\mu_i|K * \nu_j)(f) - \mu|K * \nu(f)| < \varepsilon$. But we also have $|(\mu_i|K * \nu_j)(f) - \mu_i * \nu_j(f)| \leq \varepsilon\|\nu_j\| \, \|f\|$ (see 14.5 (i) or (v) below) so the result follows. □

Further basic properties are:

14.5 PROPOSITION. *Let* G *be a locally compact group and* $\lambda, \mu \in \mathcal{M}(G)$, *with* $S(\lambda)$ *compact. Then:*

(i) *for* $f \in \mathcal{K}(G, \mathbb{R})$,

$$\lambda * \mu(f) = \int\left(\int f(xy)\,d\lambda(x)\right)d\mu(y)$$
$$= \int\left(\int f(xy)\,d\mu(y)\right)d\lambda(x),$$

(ii) $S(\lambda * \mu) \subset S(\lambda)\cdot S(\mu)$ _with equality if_ $\mu, \lambda \geq 0$,

(iii) if $\nu \in \mathcal{M}(G)$, _and_ $S(\nu)$ _is compact, then_

$$\nu * (\lambda * \mu) = (\nu * \lambda) * \mu$$

(iv) G _is commutative iff_ $\lambda * \mu = \mu * \lambda$ _for all_ λ, μ _with_ $S(\lambda)$ _compact_

(v) _if_ $\|\lambda\| < \infty$ _and_ $\|\mu\| < \infty$ _then_ $\|\lambda * \mu\| \leq \|\lambda\| \, \|\mu\|$ _with equality if_ $\lambda \geq 0$ _and_ $\mu \geq 0$.

(vi) _if_ $\mu, \nu \geq 0$ _and_ $\mu * \nu = 0$ _then_ $\mu = 0$ _or_ $\nu = 0$.

Proof. Each is a trivial verification. For example (ii) follows from 13.5 (v) and 13.13 (iv) follows by considering the Dirac measures and (v) follows from 13.4 and 13.13 (iii). Finally, (vi) is an immediate consequence of (v). □

HAAR MEASURE

Convolution is a fundamental ingredient of Fourier analysis on locally compact groups. A related concept is the Haar measure which we shall now briefly consider:

14.6 DEFINITION. Let G be a locally compact group and $\mu \in \mathcal{M}^+(G)$. We call μ a left Haar measure iff $\varepsilon_x * \mu = \mu$ for all $x \in G$, and $\mu \neq 0$. Similarly, μ is a right Haar measure if $\mu * \varepsilon_x = \mu$ for all $x \in G$.

Let $\Psi : G \to G;\ x \mapsto x^{-1}$ and for $\nu \in \mathcal{M}(G)$, let $\tilde{\nu} = \Psi(\nu)$. (Ψ is a homeomorphism and so is proper).

For $g \in G$, let $L_g : G \to G$ be defined by $L_g(h) = gh$. Then μ is a left Haar measure iff $L_g{}^*(\mu) = \mu$ for all $g \in G$. (Verification for the reader).

14.7 PROPOSITION. (i) $(\mu * \nu)^{\sim} = \tilde{\nu} * \tilde{\mu}$

(ii) μ is a left Haar measure iff $\tilde{\mu}$ is a right Haar measure

(iii) μ is a left Haar measure iff for each $\lambda \in \mathcal{M}(G)$, with compact support such that $\lambda(1) = 1$, we have $\lambda * \mu = \mu$; (Similarly for right Haar measures)

(iv) Suppose G is compact (and possesses left, and hence right, Haar measures). Then every normalized ($\|\mu\| = 1$) left Haar measure is a normalized right Haar measure and vice versa. That is,

G *has at most one left* (*or right*) *Haar measure*.

Proof. (i) and (ii) follow easily from the definitions (see 14.5 (i)) and (iii) follows at once from (separate) continuity of the convolution and the approximation theorem 12.10. For (iv), if μ is a left Haar measure, and ν a right one, then by (iii), since G is compact, we have $\mu = \nu * \mu$ and $\nu = \nu * \mu$.

Remark. We have assumed $\mu \geq 0$ for a Haar measure because if $\varepsilon_x * \mu = \mu$ for μ then μ_+ and μ_- are both Haar measures as $(\varepsilon_x * \mu)^+ = \varepsilon_x * \mu^+$ (an easy verification).

For the non-compact, non-abelian case, the left and right Haar measures are in general different.

The main theorem on Haar measures is the following:

14.8 THEOREM. *Let* G *be a locally compact group. Then* G *possesses a* (*positive*) *left Haar measure* μ *and any other left Haar measure is the product of* μ *by some constant*.

For the general case, see for example Halmos, <u>Measure Theory</u>. We shall prove one special case and sketch a proof of another. (These have particularly simple proofs and include most cases of interest).

<u>Proof of 14.8 for G abelian and compact (Markoff-Kakutani method).</u>

Let $\mathcal{M}^1 = \{\mu \in \mathcal{M}^+(E),\ \|\mu\| = 1\}$, a compact subset (since G is compact - see 12.7). For $b \in G$, let

$$\mu_n = (\varepsilon_b + \ldots + \varepsilon_{b^n})/n .$$

By compactness of $\mathcal{M}^1$, this sequence has a cluster point, say μ_b. (If G were not compact we would have to use $\{\mu\colon \|\mu\| \leq 1\}$ and then μ_b could be zero). Also we have $\varepsilon_b * \mu_b = \mu_b$. In fact, by continuity, $\varepsilon_b * \mu$ is a cluster point of $\{\varepsilon_b * \mu_n\}$; also, $\varepsilon_b * \mu_n \to \mu_n$ since $\frac{1}{n}(\varepsilon_{b^{n+1}} - \varepsilon_b) \to 0$. Given any finite family $b_1,\ldots,b_n \in G$ there is $\mu \in \mathcal{M}^1$ such that all $\varepsilon_{b_k} * \mu = \mu$. In fact choose μ_k so that $\varepsilon_{b_k} * \mu_k = \mu_k$ and let $\mu = \mu_1 * \mu_2 * \ldots * \mu_n$. Then by commutativity, $\varepsilon_{b_k} * \mu = \mu$ for all k.

For any finite subset $J \subset G$, let μ_J be an element of $\mathcal{M}^1(E)$ with $\varepsilon_b * \mu_J = \mu_J$ for all $b \in J$. The family μ_J is a net in $\mathcal{M}^1$ with J directed by inclusion. Hence there is a convergent subnet. The limit is obviously the required Haar measure. □

The general Markoff-Kakutani fixed point theorem, of which this is a special case, is given in exercise 15.2.

The next proof is somewhat specialized, and may be omitted without loss of continuity.

★<u>Proof of 14.8 for G a Lie group</u>. (The proof assumes the reader knows the basic definitions concerning orientable manifolds and Lie groups; see for example, Abraham, <u>Foundations of Mechanics</u>, Ch. II). Choose an orientation Ω_e at the identity element. If L_g denotes left translation by g, define an orientation at $g \in G$ by $\Omega_g = L_g^* \Omega_e$. Then these define a non-vanishing n-form (orientation) Ω;($n = \dim G$) on G. Clearly $L_g^*\Omega = \Omega$, or Ω is left invariant. If μ_Ω is the corresponding measure, then μ_Ω is the left invariant measure by the change of variables formula. □

(This shows that if G has a smooth (C^∞)

structure, then the Haar measure is smooth in the sense that locally it is a smooth function times Lebesgue measure).

HOMOGENEOUS SPACES

There is a useful generalization of theorem 14.8 which we now sketch. To motivate the discussion, consider the unit sphere S^{n-1} in $\mathbb{R}^n$. The rotation group $SO(n)$ acts naturally on S^{n-1}. Thus, although S^{n-1} is not a group, one expects a unique measure on S^{n-1} invariant by rotations. This is in fact true.

14.9 DEFINITION. Let E be a locally compact Hausdorff space and G a locally compact group. Let Φ be an action of G on E; that is, a continuous map $\Phi : G \times E \to E$ such that if $\Phi_g : E \to E$ is given by $\Phi_g(e) = \Phi(g,e)$, then the map $g \mapsto \Phi_g$ is a homomorphism of G into the group of homeomorphisms of E onto E. (That is, $\Phi(gh,x) = \Phi(g,\Phi(h,x))$.) We assume also that the action is transitive: for any $x, x_0 \in E$ there exists $g \in G$ such that $\Phi(g,x_0) = x$. Under these circumstances we say E is a homogeneous G-space.

Further, we say that G acts uniformly on E iff E is a uniform space with a G-invariant base of its uniformity (if U is a vicinity, it is G-invariant iff $((x,y) \in U) \Longleftrightarrow ((\Phi_g(x),\Phi_g(y)) \in U$ for all $g \in G$.)

14.10 THEOREM. *If E is a locally compact Hausdorff space, G a locally compact group which acts uniformly on E, then there is a unique measure μ on E invariant under Φ; that is, $\Phi_g^*(\mu) = \mu$ for all $g \in G$.*

Notice that the existence of Haar measures is a special case of this. Indeed the proof is similar to that for the existence of Haar measures (in fact can be derived from it).

Remark. In the case of Lie groups acting smoothly on manifolds a much simpler proof may be given following our outline for Lie groups above. Here we always have uniqueness (among smooth measures.) This applies for example to the sphere S^{n-1} under the rotation group.

DIFFUSIONS

Finally in this section we shall consider diffusions. Intuitively, a diffusion "smears out points". More precisely, maps points into measures. (See 14.12 below). For example, the heat equation in $\mathbb{R}^n$, $\frac{\partial f}{\partial t} = \Delta f$ (= Laplacian of f) defines a diffusion if we let ε_x "evolve" for a time t_0. (Of course there is a large chunk of partial differential equation theory underlying this statement; see Yosida, Functional Analysis.)

14.11 DEFINITION. Let E, F be compact spaces. A *diffusion* is a map

$$\Phi : \mathcal{M}(E) \to \mathcal{M}(F)$$

such that Φ is continuous and linear. We say Φ is *positive* iff $\mu \geq 0$ implies $\Phi(\mu) \geq 0$, and Φ *conserves mass* (or is *conservative*) iff $\Phi(\mu)(1) = \mu(1)$.

The fundamental way of obtaining a diffusion is as follows (corresponding to a point diffusing):

14.12 PROPOSITION. *Let E and F be compact.*

Let $\varphi : E \to \mathcal{M}(F)$ be a continuous map [resp. with $\varphi(x) \in \mathcal{M}^+$, $\varphi(x)(1) = 1$]. Then define a map

$$T : \mathcal{K}(F,\mathbb{R}) \to \mathcal{K}(E,\mathbb{R})$$

by $$(Tf)(y) = \varphi(y)(f) .$$

Then the dual of T ;

$$T^* : \mathcal{M}(E) = \mathcal{K}(E,\mathbb{R})' \to \mathcal{M}(F) ;$$

defined by $$T^*(\mu)\cdot f = \mu(Tf)$$

is a diffusion [resp. positive, conservative diffusion] called the diffusion generated by φ .

Furthermore, all diffusions arise this way.

Proof. Clearly T is continuous [resp. of norm one and is positive]. Thus T^* is a diffusion [resp. positive conservative diffusion].

For the converse, define $\varphi(y) = \Phi(\varepsilon_y)$ if Φ is the given diffusion, and define T^* as above. Then by definition, $T^*(\varepsilon_y) = \Phi(\varepsilon_y)$, and hence by the approximation theorem 12.10, $T^*(\mu) = \Phi(\mu)$ for all $\mu \in \mathcal{M}(E)$. □

14.13 EXAMPLES. (i) Let E, F be compact and $g : E \to F$, $\alpha : E \to \mathbb{R}$ continuous. Define the map

$$\varphi : E \to \mathcal{M}(F);\ x \mapsto \alpha(x)\varepsilon_{g(x)} .$$

Then φ induces a diffusion by 14.12.

A variant is to have maps $\alpha_k : E \to \mathbf{R}$, $g_k : E \to F$, $k = 1,\ldots,n$, and let $\varphi(x) = \Sigma\, \alpha_k(x)\, \varepsilon_{g_k(x)}$.

(ii) Let G be a compact group and $\mu \in \mathcal{M}(G)$. Then $\varphi(x) = \varepsilon_x * \mu$ defines a diffusion on G. Here $\Phi(\lambda) = \lambda * \mu$. (See above proof.

(iii) Suppose $\Phi: \mathcal{M}(E) \to \mathcal{M}(E)$ is a positive conservative diffusion and $\mathcal{M}^1 = \{\mu \in \mathcal{M}(E): \mu \geq 0,\ \|\mu\| = 1\}$. Let $X_\Phi = \{\mu \in \mathcal{M}^1(E): \Phi(\mu) = \mu\}$. Then X_Φ is a closed (and hence compact) and convex set. The Markoff-Kakutani argument above (slightly modified) shows that X is not empty, that is, measures invariant under the diffusion exist. (Iterate $\Phi^n(\mu)$ instead of the points). We leave the details of this proof to the reader. Also for a family Φ_i of diffusions we also have a measure invariant under each of them provided $\Phi_i\Phi_j = \Phi_j\Phi_i$ for all i,j. (This plays the role of G abelian).

An important special case is a <u>semi-group of diffusions</u> (the heat equation mentioned above is an

example), that is a family Φ_t, $t \in \mathbf{R}$ or $\mathbf{R}^+$, with $\Phi_{t_1+t_2} = \Phi_{t_1} \circ \Phi_{t_2}$. (For the abstract Markoff-Kakutani theorem, see problem 22.2).

PROBLEMS FOR §14

14.1 Convolutions of measures and functions

Let G be a locally compact group with right Haar measure λ. For $f, g \in \mathcal{K}(G,\mathbf{R})$ (or more generally in $L_1(G,\lambda)$), define the convolution of f and g by

$$g * f(x) = \int g(xy^{-1})f(y)d\lambda(y) .$$

Show that $(g\lambda) * (f\lambda) = (g * f)\lambda$.

14.2 Properties of convergence on convolutions

Let $\mathcal{M}^+ = \mathcal{M}^+(\mathbf{R})$ with the vague topology and $\delta = \varepsilon_0$.

(i) Let $\lambda_n, \mu_n \in \mathcal{M}^+$ be sequences with $\lambda_n(\mathbf{R}) \leq k$ and $\mu_n(\mathbf{R}) \leq k$, for a fixed constant k

(a) Show that if $\lambda_n \to \lambda$, $\mu_n \to \mu$ with $\lambda_n(\mathbf{R}) \to \lambda(\mathbf{R})$ then $\lambda_n * \mu_n \to \lambda * \mu$.

(b) Suppose $\mu, \nu \in \mathcal{M}^+$ are such that there is a unique $\lambda \in \mathcal{M}^+$ with $\lambda * \mu = \nu$. Suppose

$\lambda_n * \mu_n \to \nu$, $\mu_n \to \mu$ and either $\lambda_n(\mathbf{R}) \to \lambda(\mathbf{R})$ or $\mu_n(\mathbf{R}) \to \mu(\mathbf{R})$. Then show $\lambda_n \to \lambda$.

Apply this to the cases $\mu = \delta$ and $\mu = \nu$ and determine λ .

(ii) Let $\pi_n \in \mathcal{M}^1(\mathbf{R})$ be a sequence and let

$$\mu_n = \pi_1 * \pi_2 * \cdots * \pi_n$$

and

$$\mu_{p,q} = \pi_p * \pi_{p+1} * \cdots * \pi_q, \text{ for } p < q$$

(a) If $\mu_n \to \mu$, $\mu \in \mathcal{M}^1$, does $\pi_n \to \delta$? What about the converse? Is the condition $\mu \in \mathcal{M}^1$ crucial?

(b) Show that μ_n converges in $\mathcal{M}^+(\mathbf{R})$ iff $\lim\limits_{p,q\to\infty} \mu_{p,q} = \delta$.

(c) Give a sufficient condition for the convergence of μ_n expressing that the supports of the π_n are compact and quickly converging to 0 .

CHAPTER 4

TOPOLOGICAL VECTOR SPACES

This chapter begins our study of topological vector spaces. The first section of the chapter establishes the basic definitions. The second is devoted to the study of the space $\mathcal{K}(E,\mathbf{R})$ as a topological vector space. Perhaps the most important facts in this chapter are the open mapping theorem, the closed graph theorem and the description of a locally convex space in terms of seminorms.

Standard references for this chapter are Bourbaki, [2] (livre XV); Schaefer [1], <u>Topological</u>

Vector Spaces; Kelley-Namioka, [2] Linear Topological Spaces; and Yosida [1], Functional Analysis.

§15 BASIC PROPERTIES OF TOPOLOGICAL VECTOR SPACES

This section establishes the fundamental properties of topological vector spaces such as metrizability, completeness and the basic definitions (absorbing, circled, bounded set, etc.) Two important theorems are that finite dimensional subspaces are isomorphic to Euclidean space and that a topological vector space is finite dimensional iff it is locally compact.

15.1 DEFINITION. We let $\mathbb{K}$ denote either the reals $\mathbb{R}$ or the complex numbers $\mathbb{C}$. A topological vector space (a TVS, for short) is a vector space E over $\mathbb{K}$ with a topology such that the following maps are continuous

$$\varphi : E \times E \to E ; (x,y) \mapsto x + y$$

$$\psi : \mathbb{K} \times E \to E ; (\lambda,x) \mapsto \lambda x .$$

For example, any Banach space is a TVS. There are many important TVS's however, which are not Banachable. One of these, $\mathcal{K}(E,\mathbb{R})$, will be studied in detail in the next section.

We shall find the following convenient for checking the axioms.

15.2 PROPOSITION. Suppose E is a vector space over $\mathbb{K}$ with a topology in which addition is continuous. Then E is a topological vector space iff the maps $\lambda \mapsto \lambda x_0$ and $x \mapsto \lambda_0 x$ are continuous for each $\lambda_0 \in \mathbb{K}$, $x_0 \in E$, and the map $(\lambda, x) \mapsto \lambda x$ is continuous at $(0,0)$.

Proof. Fix $(\lambda_0, x_0) \in \mathbb{K} \times E$ and define $\psi(\lambda, x) = \lambda x$, $\psi'(\lambda, x) = (\lambda - \lambda_0, x - x_0)$, $\alpha(\lambda, x) = \lambda_0 x$, and $\beta(\lambda, x) = \lambda x_0$. Then note that $\psi = (\alpha + \beta + \psi \circ \psi') - \lambda_0 x_0$ is continuous at (λ_0, x_0). The proposition follows at once from this. □

15.3 PROPOSITION. Let E be a TVS and $U \subset E$ an open subset, $a \in E$ and $\lambda \in \mathbb{K}$, $\lambda \neq 0$. Then

(i) the map $T_a : E \to E;\ x \mapsto x + a$ is a homeomorphism of E onto E

(ii) the map $M : E \to E;\ x \mapsto \lambda x$ is a homeomorphism of E onto E

(iii) for any subset $X \subset E$, the set

$X + U = \{x + u: x \in X, u \in U\}$ <u>is open</u>

(iv) $\lambda U = \{\lambda u : u \in U\}$ <u>is open in</u> E

(v) <u>if</u> $F \subset E$ <u>is closed, then</u> λF <u>is closed</u>

(vi) <u>if</u> $F_1 \subset E$ <u>and</u> $F_2 \subset E$ <u>are compact</u>, <u>then</u> $F_1 + F_2$ <u>is compact</u>

(vii) <u>if</u> F_1 <u>is compact and</u> F_2 <u>is closed</u> <u>then</u> $F_1 + F_2$ <u>is closed</u>.

<u>Remark</u>. If F_1 and F_2 are closed, $F_1 + F_2$ need not be closed except in special circumstances (problem 15.1). For example, in $\mathbb{R}^2$, consider F_1 and F_2 as indicated schematically in figure 15.1.

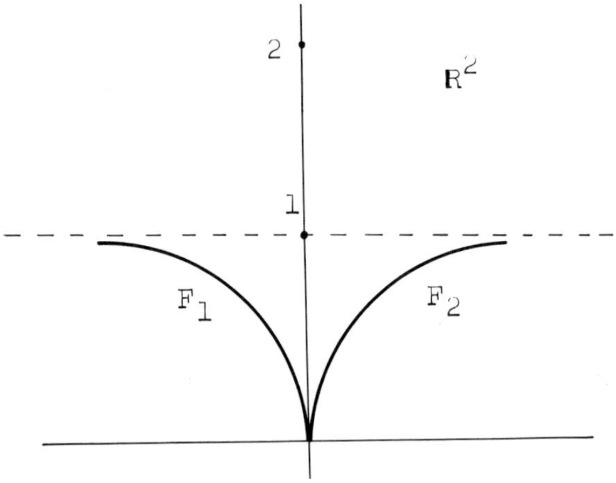

Figure 15.1

[$F_1 + F_2$ contains points arbitrarily close to (0,2), but does not contain this point].

Proof of 15.3. (i) and (ii) are obvious by continuity of addition and scalar multiplication and the fact that $(T_a)^{-1} = T_{-a}$, and $(M_\lambda)^{-1} = M_{\lambda^{-1}}$.

Then (iii) follows, as $X + U = \bigcup\{a + U: a \in X\}$ which is open by (i). (iv) and (v) are immediate by (ii).

For (vi) note that $F_1 + F_2 = \varphi(F_1 \times F_2)$ where $\varphi(x,y) = x + y$. Since $F_1 \times F_2$ is compact, so is $F_1 + F_2$.

Finally we prove (vii). Let α_i be a net in $F_1 + F_2$ and $\alpha_i \to \alpha$. Let $\alpha_i = \beta_i + \gamma_i$ where $\beta_i \in F_1$, and $\gamma_i \in F_2$. Since F_1 is compact, β_i has a convergent subnet, say $\beta_j \to \beta$. Then $\alpha_j - \beta_j \to \alpha - \beta = \gamma$ by continuity of addition. Hence $\gamma \in F_2$ and $\alpha \in F_1 + F_2$. □

Notice that by 15.3 (i), the topology is completely determined by the neighborhood base at 0.

Later we shall usually assume E to be Hausdorff, although this assumption will always be explicit. In this regard we have the following (valid more generally for topological groups):

15.4 PROPOSITION. Let E be a TVS. Then the

following are equivalent:

(i) E is Hausdorff

(ii) $\{0\}$ is closed

(iii) $\{0\} = \bigcap\{V : V$ is a neighborhood of $0\}$

(iv) E is T_0.

Proof. The following implicationsaare trivial. (i) $\Rightarrow$ (ii) $\Rightarrow$ all points are closed $\Longleftrightarrow T_1 \Rightarrow$ (iii) $\Rightarrow$ (iv). Also by translation invariance, $T_0 \Longleftrightarrow T_1$. To complete the proof, we show (iii) $\Rightarrow$ (i). Let $a \in E$, $a \neq 0$ and V be a neighborhood of 0 disjoint from a. There exists a neighborhood of 0, U, such that $U + U \subset V$ and $U = -U$ (by continuity of addition there is W so $W + W \subset V$. Let $U = W \cap (-W)$.) The proof will be complete if we show that U and $U + a$ are disjoint. Indeed, if $b \in U \cap (U + a)$, then $a = b - c$ for some $c \in U$ which implies $a \in U - U = U + U \subset V$, contradiction. □

Now we recall some definitions familiar to the reader:

15.5 DEFINITION. Let E be a vector space over K. Then $X \subset E$ is an affine subspace iff $X = Y +$

for a subspace Y and some $b \in E$.

A set $X \subset E$ is convex iff $(x,y \in X,$ $0 \leq \lambda \leq 1) \Rightarrow (\lambda x + (1-\lambda)y \in X)$.

A set $X \subset E$ is a cone, iff $x \in X$, $\lambda > 0$ implies $\lambda x \in X$ (0 may or may not belong to X). A cone is proper iff $X \cap (-X)$ is $\{0\}$ or $\emptyset$, and is called pointed if $0 \in X$.

If $F \subset E$ is a linear subspace we let E/F denote the quotient of E under the equivalence relation $x \approx y$ iff $x - y \in F$, and endow E/F with its natural vector space structure.

15.6 PROPOSITION. *Let E be a TVS and $X \subset E$. Then if X has any of the following properties (i) - (iv), so does the closure $c\ell(X)$:*

(i) X *is a subspace*

(ii) X *is an affine subspace*

(iii) X *is convex*

(iv) X *is a cone*.

Proof. Each follows easily by continuity. For example, we prove (iii). Let x_i be a net in X, $x_i \to x$, and similarly $y_j \to y$. We must show that $\lambda x + (1-\lambda)y \in c\ell(X)$; but clearly $\lambda x_i + (1-\lambda)y_j$ is a net in X converging to $\lambda x + (1-\lambda)y$. □

15.7 PROPOSITION. *Let E be a TVS and $F \subset E$ a linear subspace. Then E/F with the quotient topology is a TVS. The canonical mapping $E \to E/F$ is open. Further, E/F is Hausdorff iff F is closed.*

Proof. Let $[x]$ denote the equivalence class of $x \in E$. Let U be a neighborhood of $[x]+[y] = [x+y]$ in E/F . Then $\pi^{-1}(U)$ is a neighborhood of $x+y$ in E where $\pi : E \to E/F$ is the projection. Hence there are neighborhoods V, W of x and y such that $V+W \subset \pi^{-1}(U)$. Now for any open set $A \subset E$, $\pi(A)$ is open, for

$$\pi^{-1}(\pi(A)) = F + A$$

open, by 15.3 (iii), (i.e. the map π is open). Therefore $\pi(V)$ and $\pi(W)$ are open neighborhoods of $[x]$ and $[y]$, and $\pi(V)+\pi(W) \subset U$. Similarly, scalar multiplication is continuous.

The last part follows since F is closed iff $\{[0]\} \subset E/F$ is closed. ($F = \pi^{-1}(\{[0]\})$ and π is continuous, so $[0]$ closed implies F is closed

Conversely, if F is closed then so is $[0]$, since π is an open mapping and $\pi(E\setminus F) = (E/F)\setminus\{[0]\}$.) □

Notice that for any TVS E, $E/c\ell\{0\}$ is a Hausdorff TVS. Also, recall that if E is normable and F is closed then E/F is normable by $\|[x]\| = \inf\{\|x+y\|: y \in F\}$. We shall see below that under certain conditions completeness and metrizability are also inherited.

Another situation in which 15.7 has a role is this: Let E,F be Hausdorff TVS's and $f: E \to F$ a continuous linear map. Then $\ker(f) = \{x \in E: f(x) = 0\}$ is a closed subspace of E, so $E/\ker(f)$ is a Hausdorff TVS. Also, f induces a unique continuous linear map $\tilde{f}: E/\ker(f) \to F$ such that $f = \tilde{f} \circ \pi$ where $\pi: E \to E/\ker(f)$ is the projection.

15.8 DEFINITIONS. Let E be a TVS and $X \subset E$. We shall let $v(X)$ denote the linear subspace generated by X (that is $\{\sum_{i=1}^{n} \lambda_i x_i: \lambda_i \in K, x_i \in X\}$). We say that X is <u>total</u> iff $v(X)$ is everywhere dense in E.

A family $\{a_i\}$ of elements of E is called <u>topologically free</u> iff no a_j is contained in the closure of the space spanned by $\{a_i: i \neq j\}$.

More generally, a family $\{a_i\}_{i \in I}$ is called <u>strongly topologically free</u> iff for any $I_1, I_2 \subset I$, $I_1 \cap I_2 = \emptyset$, we have

$$c\ell(v\{a_i : i \in I_1\}) \cap c\ell(v\{a_i : i \in I_2\}) = \{0\} .$$

15.9 <u>PROPOSITION</u>. <u>Let</u> E <u>be a</u> TVS. <u>Then</u>

(i) <u>for</u> $X \subset E$, <u>we have</u> $c\ell(v(X)) \supset v(c\ell(X))$.

(ii) <u>if</u> E <u>has a total denumerable subset then</u> E <u>is separable</u>.

<u>Proof</u>. Since $c\ell(v(X))$ is closed it contains $c\ell(X)$ and since it is a vector space it contains $v(c\ell(X))$.

It is clear that if X is denumerable and $v(X)$ is dense in E, then $\{\sum_{k=1}^{n} r_k x_k : r_k$ is rational ($r_k + is_k$ in the complex case), and $x_k \in X\}$ is a countable dense subset of E by continuity of addition and scalar multiplication. □

Next we inquire about which properties of a neighborhood base of 0 will determine a topology making a vector space E into a TVS. For this we first prepare:

15.10 <u>DEFINITION</u>. Let E be a vector space over $\mathbb{K}$.

We say that $X \subset E$ is <u>circled</u> (or <u>balanced</u>) iff $\lambda \in \mathbb{K}$, and $|\lambda| \leq 1$ implies that $\lambda X \subset X$. Furthermore, X is called <u>absorbing</u> (or <u>radial</u> at 0) iff X absorbs each point $x \in E$.

If E is a TVS, a set $X \subset E$ is called <u>bounded</u> iff for each neighborhood U of 0 there is a $\lambda \in \mathbb{K}$ so that $X \subset \lambda U$.

The reader is cautioned that some authors vary the definitions of absorbing and radial.

The next theorem is extremely convenient for showing that a given space is a TVS.

15.11 <u>THEOREM</u>. *Let E be a TVS. Then there is a base $\mathfrak{B}$ of neighborhoods [which may be chosen closed or open] of 0 such that*

(i) each $V \in \mathfrak{B}$ is absorbing and circled

(ii) $V \in \mathfrak{B}$ implies $\lambda V \in \mathfrak{B}$ for each $\lambda \neq 0$, $\lambda \in \mathbb{K}$,

and (iii) for each $V \in \mathfrak{B}$ there exists $W \in \mathfrak{B}$ such that $W + W \subset V$.

Conversely, given a vector space E and a collection $\mathfrak{B}$ satisfying (i), (ii), and (iii), then there is a unique TVS structure on E for which $\mathfrak{B}$ is a base of neighborhoods of 0.

<u>Proof</u>. Let W be a neighborhood of 0. By continuity of scalar multiplication there is an $\varepsilon > 0$ and a neighborhood U of 0 so $|\lambda| \leq \varepsilon$ implies $\lambda U \subset W$. Therefore, $\bigcup\{\lambda U : |\lambda| < \varepsilon\}$ is a circled neighborhood of 0. Hence the circled neighborhoods of 0 are absorbing by continuity of scalar multiplication. Property (ii) is clear and (iii) is just continuity of addition. (Notice that we may choose the neighborhoods closed or open). This proves the first part.

By translation invariance, the topology is unique if it exists. All that remains is to prove that addition and scalar multiplication are continuous in the topology defined by $\mathfrak{B}$ satisfying (i), (ii), (iii). But (iii) says that addition is continuous at 0, so it is continuous at each point by translation invariance.

Now $(\lambda, x) \mapsto \lambda x$ is continuous at $(0,0)$ since the neighborhoods are circled. For each x_0, the map $\lambda \mapsto \lambda x_0$ is continuous since the neighborhoods are obsorbing. For each $\lambda_0 \neq 0$, the map $x \mapsto \lambda_0 x$ is continuous since it is continuous at 0 (by (ii)) and the topology is translation invariant. Thus by 15.2 we have the result. □

Bounded sets will be considered in detail in a later section. For the present, we note the following basic properties.

15.12 PROPOSITION. Let E be a TVS and $X \subset E$.

(i) X is bounded iff each neighborhood of 0 absorbs X

(ii) if X is bounded then so are (a) every subset of X, (b) $c\ell(X)$, (c) λX for $\lambda \in \mathbb{K}$,

(iii) if X and Y are bounded then so are $X \cup Y$ and $X + Y$

(iv) if E is a normed space, $X \subset E$ is bounded iff there is a constant M such that $x \in X$ implies $\|x\| \leq M$. (Thus bounded sets in a TVS generalize the notion of sets bounded in norm)

(v) any compact subset of a TVS is bounded.

Each proof follows easily and the details will be left to the reader. For example, (i) follows from the fact that circled neighborhoods form a base for the neighborhood system of 0. For (v), observe that for any circled neighborhood V of 0, $\{nV : n = 1,2,\ldots\}$ cover the compact K, so there is a constant N such that $K \subset NV$, and

K is bounded. □

We now show that each TVS is a uniform space in a natural way so that we can talk about uniform continuity, completeness etc.

15.13 PROPOSITION. *Every* TVS *is uniformizable. A base of vicinities is given by the family sets*

$$V_U = \{(x,y) \in E \times E : x - y \in U\}$$

for U *a neighborhood of* 0 .

Proof. First we check that this family of sets satisfy the properties of 5.3. First, it is a filter base for the neighborhoods of 0 are closed under finite intersections. That $\Delta \subset V_U$ is obvious. Also, if $U' \subset U$ is circled, then $V_{-U} = V_U^{-1} \supset V_{U'}$, so V_U^{-1} lies in the filter. For a given U , choose W such that $W + W \subset U$, which implies $V_W \circ V_W \subset V_U$.

Second we check that the uniform topology coincides with the original one. But the uniform topology is generated by the neighborhoods

$$V_U(x) = \{y \in E : x - y \in U\} = x - U$$

so that the assertion follows. □

15.14 <u>DEFINITION</u>. The TVS <u>category</u> has as objects the topological vector spaces and as morphisms, the uniformly continuous linear maps. An isomorphism in this category is called a TVS <u>isomorphism</u>.

Just as with uniform spaces, we can complete a TVS.

15.15 <u>PROPOSITION</u>. *Let* E *be a Hausdorff* TVS. *Then* E *has a Hausdorff* TVS *completion which is unique up to a* TVS *isomorphism*.

To prepare the proof, we first prove the following.

15.16 <u>PROPOSITION</u>. *Let* $\{E_i\}$ *be a family of* TVS's *and* E *their product*. *Then* E *is a* TVS. *If each* E_i *is complete, so is* E .

Proof. It suffices to show that E is a TVS (see 5.18). To show that addition is continuous, suppose $x_\alpha \to x$ and $y_\alpha \to y$ in E . Then the ith components also converge, $x_\alpha^i \to x^i$ and $y_\alpha^i \to y^i$ which implies $x_\alpha^i + y_\alpha^i \to x^i + y^i$ since E_i is a TVS.

This means that $x_\alpha + y_\alpha \to x + y$. Scalar multiplication is similar. □

Proof of 15.15. Let $\{d_i\}$ be the family of pseudo-metrics generating the uniformity as constructed in 5.7. By that proof, each d_i is translation invariant, that is, $d_i(x+a, y+a) = d_i(x,y)$. Using 15.11 and the fact that the d_i metric topology is coarser than the E topology, we see that (E,d_i) is a TVS. We claim that the completion of (E,d_i) as a pseudo-metric space is also a TVS. Indeed, addition is uniformly continuous, so extends uniquely to the completion as a continuous map. For each fixed λ, $x \mapsto \lambda x$ is uniformly continuous, so also extends. To show that scalar multiplication is continuous, we check 15.2. First, the map $\lambda \mapsto \lambda x_0$ is continuous because it is uniformly continuous on $\mathbb{R}$ (or $\mathbb{C}$) since $d_i(\lambda x_0, \lambda_1 x_0) = d_i((\lambda-\lambda_1)x_0, 0)$. Similarly, $x \mapsto \lambda_0 x$ is continuous. That $(\lambda, x) \mapsto \lambda x$ is continuous at 0 follows from the inequality $d_i(\lambda x, 0) \leq d_i(\lambda(x-x_0), 0) + d_i(\lambda x_0, 0)$ where we choose $x_0 \in E$ and use the fact that $(\lambda, x_0) \mapsto \lambda x_0$ is continuous. More precisely, given $\varepsilon > 0$, choose $\delta > 0$ such that

$|\lambda| < \delta$ and $d(x_o,0) < \delta$ imply that $d(\lambda x_o,0) < \varepsilon/2$ (for $x_o \in E$). Then choose $|\lambda| < \delta$ and $x \in \hat{E}$ with $d_i(x,0) < \delta/2$. Find $x_o \in E$ so that (for λ and x fixed), $d_i(\lambda x, \lambda x_o) < \varepsilon/2$ and $d_i(x,x_o) < \delta/2$ and from the above inequality, deduce that $d_i(\lambda x,0) < \varepsilon$.)

Now following the proof of 5.21, we embed E as a subspace in a product of complete TVS's (which is complete by 15.16). The closure of this subspace is the required completion. For uniqueness, observe that in 5.20, if f is linear, so is its extension. □

Regarding uniformly continuous maps, note that a map $f: E \to F$, where E,F are TVS's, is uniformly continuous iff given any neighborhood V of 0 in F, there is a neighborhood U of 0 in E such that $x-y \in U$ implies $f(x)-f(y) \in V$. For example the maps T_a and M_λ in 15.3 are TVS isomorphisms.

To determine if a TVS is metrizable we use the following important theorem.

15.17 THEOREM. *Let* E *be a Hausdorff* TVS. *Then*

the following are equivalent:

(i) *the uniformity on* E *is metrizable*

(ii) *the topology on* E *is metrizable*

(iii) E *has a denumerable base of* 0-*neighborhoods*.

Proof. Clearly (iii) $\Longleftrightarrow$ the uniformity has a bountable base $\Longleftrightarrow$ (i) by 5.9. The implications (i) $\Longrightarrow$ (ii) $\Longrightarrow$ (iii) are trivial. □

For a normed space E, if F is a closed subspace then E/F has the structure of a normed space, and is complete if E is. More generally, we have the following:

15.18 *PROPOSITION*. *Let* E *be a metrizable* TVS *and* $F \subset E$ *a subspace*. *If* F *is closed*, E/F *is metrizable*. *If* E *is complete and metrizable, and* F *is closed, then* E/F *is complete and metrizable*.

Proof. For F closed, E/F is Hausdorff (it is a TVS with {0} closed). Since the topology is the final one for $\pi : E \to E/F$, E/F has a countab[le] base of 0-neighborhoods, and so is metrizable.

Next, let E be complete metrizable. Since

E/F is metrizable, it suffices to prove that every Cauchy sequence converges. Let V_n ; $n=1,2,\ldots$ be a base of symmetric neighborhoods of 0 in E ; then $[V_n] = \pi(V_n)$ is a base of neighborhoods for E/F . Choose this base such that $V_{n+1} + V_{n+1} \subset V_n$. Let $[x_k]$ be a Cauchy sequence in E/F and select a subsequence y_k such that $[y_\ell] - [y_m] \subset V_n$ for $\ell, m \geq n$ (definition of a Cauchy sequence).

We claim that for $z \in [y_m]$, and $m, \ell \geq n$, then $[y_\ell] \cap (z + V_n) \neq \emptyset$. (Indeed, if $y \in [y_\ell]$, then $y \in z + V_n + F$ since $[y_\ell] - [y_k] \subset V_n$. If $y = z + \upsilon + u$; $\upsilon \in V_n$, $u \in F$ then $y - u = z + \upsilon$ belongs to $z + V_n$ and also to $[y_\ell]$.)

Next, define a sequence t_n in E inductively as follows. Let $t_0 \in [y_0]$, and given t_n , choose $t_{n+1} \in [y_{n+1}] \cap (t_n + V_n)$ which is possible by the above remark.

But for this sequence t_n , we have $t_{n+p} - t_n \in V_{n-1}$. (Inductively we have

$$t_{n+p} \in t_{n+p-1} + V_{n+p-1} \subset t_{n+p-2} + V_{n+p-2} + V_{n+p-1}$$
$$\subset \ldots \subset t_n + V_n + V_{n+1} + \ldots + V_{n+p-1}$$
$$\subset t_n + V_{n-1}) .$$

Therefore, t_n is a Cauchy sequence in E and hence converges to, say, $x \in E$. Therefore $[y_n] = [t_n]$ converges to $[x]$ in E/F. Thus $[x]$ is a cluster point of the original Cauchy sequence $[x_k]$ and hence $[x_k]$ converges to $[x]$. □

Remark. For an example of a complete TVS E with a closed subspace F such that E/F is not complete, see Bourbaki, livre V, ch. iv, §4, Ex. 10(b).

Sometimes a vector space has two topologies which makes it a TVS. If one of these is complete, the following gives a useful criterion for completeness of the other topology. Applications will be given later in the book.

15.19 PROPOSITION. *Let* E *be a vector space and* $\mathfrak{T}_1$, $\mathfrak{T}_2$ *two topologies making* E *a* TVS. *Suppose*

(i) $\mathfrak{T}_2 \subset \mathfrak{T}_1$ ($\mathfrak{T}_1$ *is finer than* $\mathfrak{T}_2$),

and (ii) 0 *has a base of* $\mathfrak{T}_1$ *neighborhoods which are* $\mathfrak{T}_2$*-closed*.

Then if $(E, \mathfrak{T}_2)$ *is complete, so is* $(E, \mathfrak{T}_1)$.

Proof. Let x_i be a $\mathfrak{T}_1$ Cauchy net. Since $\mathfrak{T}_2$ is coarser, x_i is also $\mathfrak{T}_2$ Cauchy and hence converges $(\mathfrak{J}_2)$, say to x. For any neighborhood V of 0 in $\mathfrak{J}_1$ which is $\mathfrak{J}_2$ closed there is an i_o such that $i,j \geq i_o$ implies $x_i - x_j \in V$. Since V is $\mathfrak{J}_2$ closed, this implies that $x_i - x \in V$ for $i \geq i_o$ which proves the assertion. □

Next we consider initial topological vector spaces which generalize the product of TVS's.

15.20 PROPOSITION. Let E be a vector space and $\{F_i\}$ a family of TVS's. Let $f_i: E \to F_i$ be a family of linear mappings. The initial topology with respect to the f_i makes E a TVS. If E is Hausdorff, then $f: E \to \Pi F_i$, with components f_i is a TVS isomorphism of E onto a subspace of ΠF_i. In particular, if the F_i are complete and Hausdorff, then $c\ell(f(E))$ is the completion of E.

Proof. To show that E is a TVS, one may proceed as in 15.16, or note that E has a base (generated by $f_i^{-1}(\mathfrak{B}_i)$) satisfying 15.11.

If E is Hausdorff and $a \neq b$ then $f(a) \neq f(b)$. Indeed, by 4.19, the topology is the initial topology with respect to f and $f(a) = f(b)$ would imply a and b could not be separated.

Also from 4.19, f is continuous and being linear is uniformly continuous, and has a continuous inverse. □

The rest of this section will be devoted to the role of dimension in a TVS. In particular, we want to show that finite dimensional TVS's are exactly the locally compact ones.

We begin with a remark on hyperplanes:

15.21 DEFINITION. Let E be a vector space over $\mathbf{K}$. A *hyperplane* in E is a subspace $H \subset E$ of codimension one (E/H is one dimensional).

15.22 PROPOSITION. (i) *Let* E *be a vector space over* $\mathbf{R}$ *and* $\ell : E \to \mathbf{R}$ *a linear map* $\ell \neq 0$. *Then* $H = \ker \ell$ *is a hyperplane. In fact all hyperplanes arise this way.*

(ii) *Let* E *be a* TVS *and* $H \subset E$ *a hyperplane. Then either* H *is closed or is everywhere dense in* E.

Proof. (i) Let $\ell : E \to \mathbf{R}$ be linear and $\ell(x_0) \neq 0$. Then let F denote the subspace spanned by x_0 and $H = \ker \ell$. We must show that

$E = F \oplus H$. In fact for $x \in E$. we have

$x = \frac{\ell(x)}{\ell(x_o)} x_o + (x - \frac{\ell(x)}{\ell(x_o)} x_o)$ the first term belonging to F , the second to H . The converse is clear for if $E = H \oplus F$ where F is generated by x_o , we can define a linear map $\ell: E \to \mathbf{R}$ by $\ell(H) = 0$, $\ell(x_o) = 1$, so that $H = \ker \ell$.

(ii) follows at once since $\bar{H}$ is a subspace and so is either H or E . □

Here we may also use $\mathbf{C}$ instead of $\mathbf{R}$.

There are two main theorems on finite dimensionality we want to prove. The first is this:

15.23 THEOREM. *Let* E *be a Hausdorff* TVS *over* $\mathbf{K}$ *having finite dimension* n . *Then* E *is* TVS *isomorphic to* $\mathbf{K}^n = \mathbf{K} \times \ldots \times \mathbf{K}$, *with the standard* (*Euclidean*) *structure*. *In particular*, E *is complete*.

15.24 LEMMA. *Let* E *be a Hausdorff* TVS *over* $\mathbf{K}$, *of dimension one*. *Let* $a \in E$, $a \neq 0$. *Then the map* $\varphi : \mathbf{K} \to E$; $\lambda \to \lambda a$ *is a* TVS *isomorphism*.

Proof. First note that φ is an algebraic isomorphism and is continuous. Given $\varepsilon > 0$ there is a circled neighborhood U of 0 in E such that $\varepsilon a \notin U$, since the topology is Hausdorff. Then $\lambda a \in U$ implies $|\lambda| < \varepsilon$ (if $|\lambda| \geq \varepsilon$ then $\varepsilon a = (\varepsilon\lambda^{-1})(\lambda a) \in U$ since $|\varepsilon \lambda^{-1}| \leq 1$ and U is circled). Hence the inverse is continuous. □

The following definitions will also be useful.

15.25 DEFINITION. Let E be a vector space over K. Subspaces F_1 and F_2 are called *algebraic supplements* iff $E = F_1 \oplus F_2$. If E is a TVS, then subspaces F_1 and F_2 are *topological supplements* iff they are algebraic supplements and E is the topological direct sum (i.e. the topology of E is that of the product $F_1 \times F_2$.

15.26 LEMMA. *Let* E *be a Hausdorff* TVS *and* H *a closed hyperplane and* $E = H \oplus F$ *where* F *is one dimensional. Then* H *and* F *are topological supplements.*

Proof. We must show that the map $\varphi : H \times F \to E$, $(x,y) \mapsto x + y$ is a homeomorphism. It is continuous by continuity of addition. The inverse is continuous provided the projections $\pi_1: E \to H$ and $\pi_2: E \to F$ are continuous. Now F is homeomorphic to E/H since H is closed and both are one dimensional Hausdorff; then $\pi_2: E \to E/H \approx F$ is continuous. For continuity of π_1, just notice that $\pi_1 = (\text{identity mapping of } E) - \pi_2$. □

Proof of 15.23. We proceed by induction of n. The case $n = 1$ is covered by 15.24. The induction is effected by writing $E = H \oplus F$ for an $n-1$ dimensional subspace, and applying 15.24, 15.26 (H is closed as it is isomorphic to $\mathbb{K}^{n-1}$ and thus is complete.) □

Notice that the TVS isomorphism here is just that associated with any basis. In particular any finite dimensional Hausdorff TVS is normable since $\mathbb{K}^n$ is normable.

The proof of 15.26 also yields the following useful observation.

15.27 PROPOSITION. _Let_ H _be a hyperplane in a Hausdorff_ TVS E. _Then_ H _is closed iff there is a continuous linear map_ $\ell: E \to \mathbb{R}$ _with_ $H = \ker \ell$.

Proof. If ℓ is continuous, H is obviously closed. Conversely, if H is closed, write $H \oplus F = E$ where $\ell(h, \lambda x_0) = \lambda$ for some fixed $x_0 \in F$. By 15.26, $H \oplus F = E$ is isomorphic (TVS) to $H \times F$ and thus ℓ is continuous. □

As another corollary, note that if $H \subset E$ is a closed subspace with codimension n and algebraic supplement F (of dimension n), then H and F are topological supplements. (The proof is the same as 15.26.) In general however, a closed subspace need not have a topological supplement.

The second main theorem is the following:

15.28 THEOREM (F. Riesz). *Let E be a Hausdorff TVS. Then E is finite dimensional iff E is locally compact.*

Proof. (due to G. Choquet) By 15.23 it is clear that if E is finite dimensional, then it is locally compact.

Conversely, let V be a compact absorbing neighborhood of 0 and write $V \subset \bigcup_{i=1}^{n} (a_i + V/2)$, a finite covering by compactness, where $a_i \in V$, $i = 1,2,\ldots,n$. Set $F = v(\{a_i\})$ the subspace spanned by $a_1, \dots, a_n$, which is closed by 15.23. Now E/F is a Hausdorff TVS. We shall be done if we can show that E/F has dimension zero. Let $\varphi : E \to E/F$ be the canonical projection and note that $\varphi(V) \subset \varphi(F) + \varphi(V/2) \subset \varphi(V)/2$ since $\varphi(F) = \{0\}$. Hence we have a decreasing chain of compacts:

$$\varphi(V) \supset 2\,\varphi(V) \supset \dots \supset 2^n \varphi(V) \supset \dots$$

Since $\varphi(V)$ is absorbing, this implies that $\varphi(V) = E/F$. But E/F is compact iff its dimension is zero, by 15.23. □

PROBLEMS FOR §15

15.1 Closed subsets of $\mathcal{M}(E)$

Let E be a compact Hausdorff space and let $\mathcal{M}(E)$ have the vague topology (see §12). If F_1, $F_2 \subset \mathcal{M}^+(E)$ are closed subsets then $F_1 + F_2$ is closed. What if E is only locally compact?

15.2 The Markoff-Kakutani Theorem

Prove the general theorem of Markoff Kakutani: Let E be a Hausdorff TVS over the field $\mathbb{R}$, and K a non-empty compact convex subset of E, $\{\Gamma_i\}$ a commuting family of linear maps of E into itself, whose restrictions to K are continuous and map K into itself. Then there exists $x_0 \in K$ such that $\Gamma_i(x_0) = x_0$ for all Γ_i .

Remark. A related but deeper theorem which has important applications in partial differential equations is the Schauder-Tychonoff fixed point theorem, which is the following: Let E be a Hausdorff locally convex space (see §21) and $K \subset E$ a convex compact subset of E and $f: K \to K$ a continuous map. Then f has a fixed point $x_0 \in K$; i.e. $f(x_0) = x_0$.

The importance of this is that f can be a non-linear mapping. For a proof, see Edwards [1], Functional Analysis, p. 161.

15.3 Positive linear forms on a quotient space.

Let E be a compact Hausdorff topological space, A a closed subset of E , and $D(A) = \mathcal{C}(E,\mathbb{R})/\sim$ where $f \sim g$ iff there is an open neighborhood of A on which $f = g$. Let $\varphi: D(A) \to \mathcal{C}(A,\mathbb{R})$ be the obvious canonical mapping. Show that:

(i) $D(A)$ is an ordered vector space with order $[f] \leq [g]$ iff $f \leq g$ on some neighborhood of A, where $[f]$ is the equivalence class of $f \in \mathcal{C}(E,\mathbf{R})$.

(ii) If T is a positive linear form on $D(A)$, then $(\varphi(u) = \varphi(v)) \Rightarrow T(u) = T(v))$ for $u,v \in D(A)$.

(iii) Deduce from (ii) a simple representation of positive linear forms on $D(A)$.

(iv) Let $O(A)$ be the vector subspace of $D(A)$ defined by $O(A) = \ker \varphi = \{u \in D(A): \varphi(u) = 0\}$. Show that for every $v \in O^+(A)$, there is $w \in O^+(A)$ such that for <u>every</u> n, $n = 1,2,\ldots,$ $nv \leq w$. Conclude that every positive linear form on $O(A)$ is zero.

(v) Let $S = \mathbf{R}^{\mathbf{N}}/\sim$ where $(a_n) \sim (b_n)$ iff there exists N so that $n \geq N$ implies $a_n = b_n$.

Let B denote the subspace of S represented by bounded sequences. Show that any positive linear form on S is zero, but not for B. Do you see any connection with ultrafilters?

15.4 <u>A space with no continuous linear forms</u>.

Let $E = \mathcal{C}([0,1],\mathbf{R})$ and let $p : E \to \mathbf{R}$ be defined by $p(f) = \inf\{\varepsilon > 0: |f(x)| \leq \varepsilon$ for $x \in [0,1]\setminus B$ with $m(B) = \varepsilon\}$ where m denotes Lebesgue measure, and B is some Lebesgue measurable subset of $[0,1]$.

(i) Show that $d(f,g) = p(f-g)$ is a metric on E making E into a TVS (warning: p is not a norm).

(ii) Show that for any $k > 0$, the convex envelope of the ball $B_k = \{f \in E\colon p(f) \leq k\}$ is the whole space E.

(iii) Deduce from (ii) that the only linear continuous form on E is zero.

§16 THE SPACE $\mathcal{K}(E,\mathbf{R})$

The spaces $\mathcal{K}(E,\mathbf{R})$ and $\mathcal{K}^\infty(E,\mathbf{R})$ (the latter for $E = \mathbf{R}^n$ or a manifold) are of fundamental importance, for when given the inductive limit topology they have as duals the Radon measures and the distributions, respectively. In this section we study the topological properties of $\mathcal{K}(E,\mathbf{R})$ which are also shared by $\mathcal{K}^\infty(E,\mathbf{R})$.

First we recall the associated space $\mathcal{C}(E,\mathbf{R})$.

16.1 <u>DEFINITION</u>. Let E be a locally compact Hausdorff space and $\mathcal{C} = \mathcal{C}(E,\mathbf{R})$ the space of continuous maps $f\colon E \to \mathbf{R}$. For each compact $K \subset E$, let

$p_K: \mathcal{C} \to \mathbf{R}$ be defined by $p_K(F) = \sup\{|f(x)|: x \in K\}$ and consider the associated pseudo-metric $d_K(f,g) = p_K(f-g)$.

16.2 THEOREM. *If E is a σ-compact locally compact Hausdorff space, then $\mathcal{C}(E,\mathbf{R})$ is a Hausdorff* TVS *when endowed with the topology generated by the semi-norms p_K. Furthermore, $\mathcal{C}$ is complete, metrizable and $c\ell(\mathcal{K}(E,\mathbf{R})) = \mathcal{C}(E,\mathbf{R})$.* (*Recall that $\mathcal{K}$ consists of elements of $\mathcal{C}$ with compact support*).

Proof. That each p_K is a semi-norm is an easy and standard verification. Thus $\mathcal{C}$ with the topology generated by each p_K is a TVS (a semi-normed space). Therefore the topology generated by all the p_K makes $\mathcal{C}$ into a TVS. (One can also use 15.11 here). In fact, as we shall see in §19, such a family of semi-norms characterizes a locally convex space. The space $\mathcal{C}$ is Hausdorff since $f \neq 0$ implies $p_K(f) \neq 0$ for some K. (See 5.10).

To see that $\mathcal{C}$ is complete, let f_i be a Cauchy net. Then f_i converges uniformly on every compact. The limit is therefore a continuous function, so $f_i \to f$ in $\mathcal{C}$.

For metrizability, it suffices, by 5.9 to show that the topology is generated by a countable family of the p_K. Indeed, write $E = \bigcup \{K_n\}$ for $K_n \subset \operatorname{int}(K_{n+1})$. Then obviously the p_{K_n} generates the same topology.

To see that $\mathcal{K}$ is dense in $\mathcal{C}$ we must show that for any $f \in \mathcal{C}$, K compact and $\varepsilon > 0$ there is a $g \in \mathcal{K}$ such that $p_K(f - g) < \varepsilon$. Indeed, there exists $h \in \mathcal{K}$ such that $h = 1$ on K by Urysohn's lemma applied to a compact K_n such that $K \subset \operatorname{int}(K_n)$. Let $g = hf$, so that $p_K(f-g) = 0$. □

The above theorem is also true for $\mathcal{C}^k(\mathbb{R}^n, \mathbb{R})$ (or on a manifold) with $0 \leq k \leq \infty$ and semi-norms $p_K^\ell(f) = \sup\{|f(x)|, |Df(x)|, \ldots, |D^\ell f(x)| : x \in K\}$ for $0 \leq \ell \leq k$ and K compact ($\ell < k$ if $k = +\infty$). Here, $D^n f$ denotes the nth derivative. The proof involves a use of Taylor's theorem. See Abraham-Robbin, Transversal Mappings and Flows, ch. 2.

The basic difference between $\mathcal{C}$ and $\mathcal{K}$ is this. The above topology on $\mathcal{C}$ (the topology of uniform convergence on compacts, also called the compact open topology) makes $\mathcal{C}$ the projective limit of the spaces $\mathcal{C}(K, \mathbb{R})$ for $K \subset E$ compact (exercise 16.1) On the other hand, the useful way to topologize $\mathcal{K}(E,\mathbb{R})$ is as the <u>inductive limit</u> of the spaces $\mathcal{K}_K(E,\mathbb{R}) = \{f \in \mathcal{K} : S(f) \subset K\}$.

16.3 <u>DEFINITION</u>. Let E be a locally compact Hausdorff space and $\mathcal{K} = \mathcal{K}(E,\mathbb{R}) = \{f \in \mathcal{C}(E,\mathbb{R}): S(f)$ is compact$\}$. Also, let $\mathcal{K}_K(E,\mathbb{R}) = \{f \in \mathcal{K}: S(f) \subset K\}$ where $S(f)$ denotes the support of f . On $\mathcal{K}$ put the final topology for the inclusion maps $i_K : \mathcal{K}_K \to \mathcal{K}$ for each K compact and $\mathcal{K}_K$ equipped with the uniform (supremum) norm (which makes $\mathcal{K}_K$ into a Banach space).

If E is σ-compact and locally compact, we may find compacts K_n such that $E = \bigcup K_n$ with $K_n \subset \operatorname{int}(K_{n+1})$ and then this topology is equal to the one induced by the maps $i_n: \mathcal{K}_{K_n} \to \mathcal{K}$.

Thus a set $U \subset \mathcal{K}$ is a neighborhood of 0 in $\mathcal{K}$ iff for each compact K , $i_K^{-1}(U) = \mathcal{K}_K \cap U$ is a neighborhood of 0 in $\mathcal{K}_K$.

For E σ-compact, $\mathcal{K}$ is an example of an LF-<u>space</u>, or a countable <u>strict</u> inductive limit of Frechet spaces (in this case Banach spaces). The <u>strict</u> here refers to the fact that if $K_1 \subset K_2$ then the topology on $\mathcal{K}_{K_1}$ agrees with that of $\mathcal{K}_{K_2}$

when restricted to K_1 .

We want to show that $\mathcal{K}$ is complete, but is not in general metrizable. This shows that the inductive topology is quite different from the projective topology which makes $\mathcal{K}$ metrizable but not complete). To this end, we prepare the following:

16.4 PROPOSITION. *Let* E *be a locally compact and* σ-*compact Hausdorff space and* $P = \{f: E \to \mathbb{R}: f$ *is continuous and* $f(x) > 0$ *for all* $x \in E\}$. *For each* $h \in P$, *let*

$$V(h) = \{f \in \mathcal{K}(E,\mathbb{R}): |f(x)| \leq h(x) \text{ for all } x \in E\}.$$

Then the sets $V(h)$ *comprise a fundamental system of neighborhoods of* 0 (*in the inductive topology*).

Proof. Certainly $V(h)$ is a neighborhood of 0 since for each compact K , we have

$$\mathcal{K}_K \cap V(h) \supset \{g \in \mathcal{K}_K : p_K(g) = \|g\| \leq \varepsilon\}$$

where $\varepsilon = \inf\{h(x): x \in K\}$, and $\{g \in \mathcal{K}_K: \|g\| \leq \varepsilon\}$ is a neighborhood of 0 in $\mathcal{K}_K$ as $\varepsilon > 0$.

Conversely, we write $E = \bigcup K_n$ where $K_n \subset \text{int}(K_{n+1})$ and let U be a neighborhood of 0

in $\mathcal{K}$, so that $U \cap \mathcal{K}_{K_n}$ is a neighborhood of 0 in $\mathcal{K}_{K_n}$. Thus there is an ε_n such that $U \cap \mathcal{K}_{K_n}$ contains the ε_n ball about 0. We may assume that $\varepsilon_{n+1} \leq \varepsilon_n$. Choose $h_n \in \mathcal{K}^+$ by Urysohn's lemma such that $S(h_n) \subset K_{n+1}$ and $2h_n = \varepsilon_{n+1}$ on K_n and $0 \leq 2h_n \leq \varepsilon_{n+1}$ on all of E. Let $h = \sup h_n$. This sup is finite on each K_n (on K_n, $h = \sup(h_n, h_{n+1})$), so h is continuous and strictly positive. Also, $h < \varepsilon_n$ on K_n, so that $V(h) \subset U$. □

Notice that $V(h)$ is not a neighborhood of 0 in $\mathcal{C}(E,\mathbf{R})$ (or $\mathcal{K}(E,\mathbf{R})$ with the projective topology). This is because neighborhoods of 0 in $\mathcal{C}(E,\mathbf{R})$ are sets of the form $V_{K,\varepsilon} = \{f \in \mathcal{C}: \|f\| < \varepsilon$ on $K\}$, and f may be large outside K. This means that the inductive topology on $\mathcal{K}$ is strictly stronger (finer) than the projective topology, that is, the relative topology inherited from $\mathcal{C}(E,\mathbf{R})$.

Bounded sets of $\mathcal{K}$ are described as follows.

16.5 PROPOSITION. _Let_ E _be locally compact, Hausdorff, and_ σ-_compact_. _A set_ $B \subset \mathcal{K}(E,\mathbf{R})$ _is bounded_ (_see 15.10 and 15.12_) _iff there is a compact_

$K \subset E$ *and a constant* M *such that* $B \subset \mathcal{K}_K$ *and* $\|f\| \leq M$ *for all* $f \in B$.

Proof. In view of 15.12 (iv), it suffices to show that there is a compact K such that $S(f) \subset K$ for all $f \in B$ for B a bounded set. We may assume that E is not compact. Let E be the increasing union of compacts K_n such that $K_n \subset \text{int}(K_{n+1})$. If our claim is not true, there are points $x_n \in K_n \setminus K_{n-1}$ and $f_n \in B$ such that $f_n(x_n) \neq 0$. Now let $V = \{g \in \mathcal{K} : S(g) \subset K_n$ implies $\|g\| \leq f_n(x_n)/n\}$. Then V is a neighborhood of 0 in $\mathcal{K}$ (consider the inverse image in each $\mathcal{K}_K$ noting that $K \subset K_N$ for some N). However, V cannot absorb B , so B is not bounded, a contradiction. □

16.6 *COROLLARY*. *Under the hypotheses of* 16.5, *a sequence* $f_n \to f$ *in* $\mathcal{K}(E,\mathbf{R})$ *iff there is a compact* K *such that* $S(f_n) \subset K$ *for all* n *and* $\|f_n - f\| \to 0$.

Proof. By 15.12, convergent sequences are bounded sets, so the result follows from 16.5. □

Similar results hold for $\mathcal{K}^k(E,\mathbb{R})$, $0 \leq k \leq \infty$.

This corollary implies at once that $\mathcal{K}$ is sequentially complete. It is in fact complete, although a different method is required (for a Cauchy net need not be a bounded set). For example a net associated with the filter of neighborhoods of 0 converges to 0 but is certainly not bounded, by 16.4 and 16.5.

In the next theorem we prove that $\mathcal{K}(E,\mathbb{R})$ is complete but is not generally metrizable.

16.7 THEOREM. *Let* E *be locally compact, Hausdorff, and* σ-*compact. Then*

(i) $\mathcal{K}(E,\mathbb{R})$ *is complete*

and (ii) $\mathcal{K}(E,\mathbb{R})$ *is metrizable iff* E *is compact*.

Proof. First we prove completeness. We may suppose E is not compact. Let f_i be a Cauchy net. Since the inclusion $i: \mathcal{K} \to \mathcal{C}$ is continuous, f_i is a Cauchy net in $\mathcal{C}$ and so converges uniformly on every compact to some $f \in \mathcal{C}$. It will suffice to prove that $f \in \mathcal{K}$, so suppose not. Then there is a sequence of compacts $K_n \subset \operatorname{int} K_{n+1}$ and $x_n \in K_n \setminus K_{n-1}$ for which $f(x_n) \neq 0$ and $E = \bigcup K_n$.

Consider the neighborhood of 0 defined by

$V = \{g \in \mathcal{K} : g \in \mathcal{K}_{K_n}$ implies $\|g\| \leq f(x_n)/2\}$.

Now there is an i_o such that $i,j \geq i_o$ implies $f_i - f_j \in V$. Since each f_i has compact support, there is an n such that $f_{i_o}(x_n) = 0$. Thus $i \geq i_o$ implies that $f_i(x_n) \leq f(x_n)/2$, contradicting the fact that $f_i \to f$ pointwise.

Secondly, we prove the metrizability statement. First, if E is compact, $\mathcal{K}$ is obviously metrizable (it is even a Banach space). Secondly, suppose E is not compact and $\mathcal{K}$ is metrizable. Then there is a countable base $V(h_n)$ for $h_n \in P$ as in 16.4. Choose $x_n \to \infty$, $(x_n \in K_n \setminus K_{n-1})$, and let $V = \{g \in \mathcal{K} : S(g) \subset K_n$ implies $\|g\| \leq h_n(x_n)/2\}$. But V is a neighborhood of 0 containing no $V(h_n)$, a contradiction. □

Actually, completeness holds in any LF space as we shall see in §20. The above theorem also holds for the Schwartz space $\mathcal{D}(E) = \mathcal{K}^{\infty}(E,\mathbb{R})$, for E a manifold or $\mathbb{R}^n$.

There is another instructive way to see that $\mathcal{K}(E,\mathbb{R})$ is not in general metrizable, namely by showing that it is not a Baire space. (Recall that, by the Baire category theorem a complete metrizable space is Baire). This is done in the next proposition.

16.8 PROPOSITION. Let E be a non-compact, locally compact, σ-compact Hausdorff space. Then $\mathcal{K}(E,\mathbb{R})$ is not a Baire space.

Proof. We shall show that in fact $\mathcal{K}(E,\mathbb{R})$ is first category. Let $E = \bigcup K_n$ where $K_n \subset \mathrm{int}(K_{n+1})$, and note that $\mathcal{K} = \bigcup \mathcal{K}_{K_n}$. The subsets $\mathcal{K}_{K_n}$ are closed in (they are complete because the relative topology coincides with the norm topology which is complete, see problem 16.2). Moreover $\mathcal{K}_{K_n}$ is nowhere dense. To see this we show that its (open) complement is dense. Let $f \in \mathcal{K}$ and choose $x \notin S(f) \cup K_n$. By Urysohn there exists $h \in \mathcal{K}$ with $h(x) = 1$. Let $g_n = f + h/n \notin \mathcal{K}_{K_n}$. But $g_n \to f$ proving our claim, and hence the proposition. □

We conclude this section with the result that $\mathcal{M}(E)$ is usually not metrizable. We have already seen in §12 that $\mathcal{M}(E)$ is not complete even though $\mathcal{M}^+(E)$ is complete and metrizable (if E is second countable).

16.9 THEOREM. *Let* E *be a* 2nd *countable locally compact Hausdorff space. Then* $\mathcal{M}(E)$ *is not metrizable (in the vague topology) iff* E *contains an infinite compact set.*

Remark. It is easy to see that a first countable Hausdorff space is discrete iff every compact set is finite.

For the proof, we use the following lemma. In it we assume the Hahn-Banach theorem which will be proven in §21.

16.10 LEMMA. *Let* E *be a Hausdorff topological vector space with a base of convex neighborhoods of* 0 . *Let* E' *be the dual (the space of continuous linear maps* $\ell : E \to \mathbb{R}$) *with the weak topology (the initial topology for the maps* $\ell \mapsto \ell(e)$ *for each* $e \in E$. *Then* E' *is metrizable iff* E *has countable algebraic dimension.*

Proof. A base of neighborhoods of E' is given by the sets

$$\{\ell : |\ell(x_i)| < \varepsilon \ ; \ x_i \in E, \ i = 1,\ldots,n\}$$

for all n and $\varepsilon > 0$. If E has countable dimension we may select a Hamel basis for the x_i and use $\varepsilon = 1/n$ to obtain a countable base of 0-neighborhoods.

Conversely, if the x_i can be chosen from a countable set, they must span all of E. For if not, let y lie outside this span. Consider the set $V = \{\ell: |\ell(y)| < 1\}$. This is a neighborhood of 0 so contains a set $U = \{\ell \in E': |\ell(x_i)| < \varepsilon$ $i = 1,\ldots,n\}$. If y is not in the span of $x_1,\ldots,x_n$, there exists (by the Hahn-Banach theorem an $\ell \in E'$ such that $\ell(x_i) = 0$, $i = 1,\ldots,n$ and $\ell(y) = 2$. But $\ell \notin V$ and $\ell \in U$ a contradiction. □

<u>Proof of 16.9</u>. By 16.10, $\mathcal{M}(E)$ is metrizable iff $\mathcal{K}(E,\mathbf{R})$ has countable (algebraic) dimension. If E has no infinite compact sets, E is countable with the discrete topology. Then $\mathcal{K}(E,\mathbf{R})$ has countable dimension, a basis being the characteristic functions of the points.

Conversely, let $K_o \subset E$ be a infinite compact set, with $K_o \subset \text{int}(K)$, K compact. Then $\mathcal{K}_K$ either has finite dimension or uncountable dimension since it is a Banach space (exercise 8.3). It thus suffices to show that $\mathcal{K}_K$ is not finite dimensional. Suppose it is in fact of dimension n. By normality, there exists $n+1$ pairwise disjoint non empty open sets $U_1, \ldots, U_{n+1} \subset \operatorname{int}(K)$. By Urysohn's lemma there exists $f_k \in \mathcal{K}_K$ with $S(f_k) \subset U_k$ and $f_k \neq 0$. Obviously $f_1, \ldots, f_{n+1}$ are linearly independent, a contradiction. □

PROBLEMS FOR §16

16.1 Projective and inductive limits

(i) Prove that if E is a locally compact Hausdorff space, then $\mathcal{C}(E,\mathbf{R})$ is the projective limit of the spaces $\mathcal{C}(K,\mathbf{R})$, while $\mathcal{K}(E,\mathbf{R})$ is the inductive limit of the spaces $\mathcal{K}_K(E,\mathbf{R})$.

(ii) Describe the bounded sets of $\mathcal{C}(E,\mathbf{R})$.

16.2 The relative and norm topologies

Let E be a locally compact σ-compact Hausdorff space. Show that the relative topology on $\mathcal{K}_K(E,\mathbf{R})$ regarded as a subset of $\mathcal{K}(E,\mathbf{R})$ coincides with the norm topology, for $K \subset E$ compact.

§17 COMPACT OPERATORS

This section contains a brief account of the basic properties of compact mappings. The reader wishing to pursue this subject further should read Yosida [1], Functional Analysis, Ch. X, and Schaefer, [1] Topological Vector Spaces, Ch. III. Historically, the theory of compact mappings arose in the theory of integral equations around the time of Hilbert (1900-1910).

We begin with the definition:

17.1 DEFINITION. Let E and F be Hausdorff TVS's (real or complex) and $f: E \to F$ a linear mapping. We say that f is compact iff there exists a neighborhood V of 0 in E such that $c\ell(f(V))$ is compact. (Sometimes compact mappings are called completely continuous).

In the case of normed spaces the compactness condition becomes the following.

17.2 PROPOSITION. If E and F are normed spaces, $f : E \to F$ is compact iff $c\ell(f(B))$ is compact, where B is the unit ball in E, iff for every sequence x_n in E, with $\|x_n\|$ bounded, $f(x_n)$

contains a convergent subsequence (*in the norm topology on* F).

This proposition follows at once.

We also have the following, which is often included as part of the definition of compactness.

17.3 *PROPOSITION*. *Let* E *and* F *be Hausdorff TVS's and* $f: E \to F$ *be compact*. *Then* f *is continuous*.

Proof. Let U be a neighborhood of 0 in F and V a neighborhood of 0 in E such that $f(V)$ is relatively compact. Then since $c\ell(f(V))$ is compact it is bounded and hence is absorbed by U. Thus there is a k so that $|\lambda| \leq k$ implies $\lambda\, c\ell(f(V)) \subset U$. Therefore, $f(kV) \subset U$ which proves that f is continuous at 0. Since f is linear, it is continuous. □

(The converse of 17.3 is of course false. For example, consider the identity map on any Hausdorff TVS of infinite dimension.) One may also define a compact non-linear mapping as in 17.1, but then 17.3 would not hold.

The algebraic structure of the family of compact mappings is given in the next proposition.

17.4 PROPOSITION. *Let* E, F *be Hausdorff* TVS's *and let* $C(E,F)$ *denote the set of compact maps from* E *to* F. *Then* $C(E,F)$ *is a vector space. Furthermore, if* $f \in C(E,F)$ *and* $g: A \to E$ *is continuous linear, for* A *a Hausdorff* TVS, *then* $f \circ g$ *is compact. Similarly if* $h: f \to A$ *is continuous linear, then* $h \circ f \in C(E,A)$.

Proof. The first part follows at once from the definition 17.1. For the second, $g^{-1}(V)$ is the required neighborhood in A , if $c\ell(f(V))$ is compact. Finally $c\ell(h(f(V))) = h(c\ell(f(V)))$ is compact since h is continuous. □

One of the more fundamental properties is the next theorem, in which we restrict ourselves to normed spaces (for a slightly more general version, see Schaefer, p. 110).

17.5 THEOREM. *Let* E *and* F *be normed spaces with* F *complete. Let* $\mathcal{L}(E,F)$ *denote the vector space of continuous linear maps from* E *of* F *with the uniform norm. Then* $C(E,F)$ *is closed in* $\mathcal{L}(E,F)$.

Remark. Recall that $\mathcal{L}(E,F)$ is a Banach space with norm $\|f\| = \sup\{\|f(x)\|: \|x\| \leq 1\}$. Then 17.5 asserts that if f_n are compact and $f_n \to f$ in $\mathcal{L}(E,F)$ then f is compact.

Proof. Let B_E denote the unit ball in E and suppose $f_n \to f$ in norm with each f_n compact. As F is complete it suffices to show that $f(B)$ is precompact by 5.24. Given $\varepsilon > 0$ choose N so that $n \geq N$ implies $\|f - f_n\| < \varepsilon/2$, and choose $x_1, \ldots, x_m \in B_F$ with $f_N(B_E) \subset \bigcup\{x_k + (\varepsilon/2)B_F: k = 1,\ldots,m\}$. Then we have

$$f(B_E) \subset \bigcup\{(x_k + (\varepsilon/2)B_F + (f-f_N)(B_E): k = 1,\ldots,m\}$$

$$\subset \{x_k + \varepsilon B_F: k = 1,\ldots,m\},$$

proving the result. □

17.6 COROLLARY. *Let E and F be normed spaces with F complete. Suppose $f_n: E \to F$ is a continuous linear map of finite rank (that is, $f_n(E)$ is finite dimensional) for $n = 1,2,\ldots$ and $f_n \to f$ in $\mathcal{L}(E,F)$. Then f is compact.*

Proof. Since $f_n(E)$ is finite dimensional, it is locally compact. Hence f_n is compact. Thus 17.5 gives the result. □

The converse to this corollary is unknown; that is, is every compact map the limit of linear maps of finite rank? For a detailed discussion of this problem see Schaefer, [1] Ch. III, §9. In the non linear case the converse is true. See problem 17.1.

A useful application of this is to linear maps derived from a kernel:

17.7 PROPOSITION. *Let* A *and* B *be compact Hausdorff spaces and* $K \in \mathcal{C}(A \times B, \mathbb{R})$. *For* $\mu \in \mathcal{M}(B)$, *define the map*

$$T: \mathcal{C}(B,\mathbb{R}) \to \mathcal{C}(A,\mathbb{R}) \text{ by } T(\varphi)(x) = \int K(x,y)\varphi(y)d\mu(y)$$

Then T *is a compact operator (using the uniform norm on* $\mathcal{C}$*);* K *is called the kernel of* T .

Proof. By uniform continuity, it follows easily that $T(\varphi) \in \mathcal{C}(A,\mathbb{R})$ and obviously T is linear. By the Stone-Weierstrass theorem, we may write $K = \lim_{n\to\infty} K_n$ where $K_n = \sum_{k=1}^{N(n)} \alpha_k^n \otimes \beta_k^n$ for

$\alpha_k^n \in \mathcal{C}(A,\mathbf{R})$, $\beta_k^n \in \mathcal{C}(B,\mathbf{R})$ (see 13.8). Then if T_n is the mapping with K_n as kernel,

$$T_n(\varphi)(x) = \sum_{k=1}^{N(n)} \alpha_k^n(x) \int \beta_k^n(y)d\mu(y)$$

so that T_n is of finite rank, and hence is compact. But $T_n \to T$ is the operator norm since

$$\|T_n(\varphi) - T(\varphi)\| \leq \|K - K_n\| \, \|\varphi\| \, \|\mu\|.$$

Hence T is compact by 17.6. □

A related classical result is the following:

17.8 PROPOSITION. *Let* E *and* F *be locally compact Hausdorff spaces and* $\mu \in \mathcal{M}^+(E)$, $\nu \in \mathcal{M}^+(F)$. *For* $K \in L_2(E \times F, \mu \otimes \nu)$, *define*

$$T: L_2(E,\mu) \to L_2(F,\nu)$$

by

$$T\varphi(y) = \int K(x,y)\varphi(x)d\mu(x) .$$

Then T *is well defined and is a compact operator, called a Hilbert-Schmidt operator*.

Proof. First we show that $T\varphi \in L_2(F,\nu)$. Now

$$\int (\int K(x,y)\varphi(x)d\mu(x))^2 d\nu(y)$$
$$\leq \int\{\int K(x,y)^2 d\mu(x)\}\{\int\varphi(x)^2 d\mu(x)\}d\nu(y)$$

by the Schwartz inequality. Therefore

$$\|T\varphi\|_2^2 \leq \|\varphi\|_2^2 \int\int K(x,y)^2 d\mu(x)d\nu(y)$$
$$= \|\varphi\|_2^2 \|K\|_2^2$$

by Fubini's theorem.

Since $\mathcal{K}(E,\mathbf{R}) \otimes \mathcal{K}(F,\mathbf{R})$ is dense in $\mathcal{K}(E \times F,\mathbf{R})$ which is dense in L_2, we can write as in 17.7, $K_n \to K$ in $L_2(E \times F, \mu \otimes \nu)$ where K_n is of finite rank. The corresponding T_n converges to T in operator norm since

$$\|T_n(\varphi) - T(\varphi)\|_2 \leq \|K-K_n\|_2 \|\varphi\|_2$$

by the above inequality. □

This proof is a good example of the economy of using the criterion of 17.6. Direct proofs of the above Hilbert-Schmidt theorem are possible but are more complicated.

Another important class of compact operators are the nuclear operators:

17.9 DEFINITION. Let E and F be Banach spaces and $T : E \to F$ a linear map. We call T *nuclear* iff there are sequences $\alpha_n \in E'$ (the continuous linear maps of E into $\mathbb{R}$), $y_n \in F$ and $c_n \in \mathbb{R}$ with $\sup\|\alpha_n\| < \infty$, $\sup\|y_n\| < \infty$, $\Sigma|c_n| < \infty$ so that

$$T(x) = \lim_{m\to\infty} \sum_{n=1}^{m} c_n \alpha_n(x) y_n$$

for each $x \in E$.

17.10 PROPOSITION. *Nuclear operators are compact.*

Proof. Clearly T is the limit of operators of finite rank. □

For further discussion and examples, see Schaefer and Yosida.

One of the nice properties of compact operators is that they have discrete spectra. We shall confine ourselves to proving that the multiplicity of each eigenvalue is finite:

17.11 PROPOSITION. *Let* E *be a Hausdorff* TVS *and* $f : E \to E$ *compact. For each* $k \neq 0$, $k \in \mathbb{K}$

$\mathbb{K} = \mathbb{R}$ or $\mathbb{C}$), <u>the eigenspace</u> $E_k = \{x \in E: f(x) = kx\}$ <u>is finite dimensional</u>.

<u>Proof</u>. Obviously E_k is a subspace of E and $f(E_k) \subset E_k$. In fact, since $k \neq 0$, $f(E_k) = E_k$ and f is an isomorphism on E_k. From the definition of E_k and continuity of f, E_k is closed. Therefore, f restricted to E_k is a compact mapping. Let V be a neighborhood of 0 such that $c\ell(f(V))$ is compact in E_k. From the definition of f on E_k, f is an open map (in fact a homeomorphism), and hence $c\ell(f(V))$ is a compact neighborhood of 0 in E_k. Thus E_k is locally compact and hence is finite dimensional by Riesz' theorem. □

PROBLEMS FOR §17

17.1 <u>Mappings of finite rank</u>

Let E and F be normed spaces and $f: E \to F$ a continuous compact mapping (not necessarily linear). Choose $U \subset E$ such that $f(U)$ is relatively compact. Show that there exist mappings $f_n: U \to F$ each having image in some finite dimensional subspace and such that $f_n \to f$ uniformly on U.

Remark. This shows that the finite rank problem can be solved using non-linear maps.

17.2 Schauder's Theorem

Let E and F be Banach spaces and E', F' their duals with the strong (uniform) topology. Let $T: E \to F$ be a linear map and $T': F' \to E'$ its dual $(T'(\alpha))(e) = \alpha(T(E))$. Show that if T is compact then T' is compact. Apply this to the examples 17.7 and 17.8.

17.3 Compact Operators on $\mathbf{R}^I$.

(i) Let I be a set and $\{a_i\}_{i\in I}$ a family of positive numbers. Define $f: \mathbf{R}^I \to \mathbf{R}^I$ by $(f(x))_i = a_i x_i$. Show that f is continuous ($\mathbf{R}^I$ given the product topology).

(ii) Let $E = \{x \in \mathbf{R}^I: \sum_{i\in I} x_i^2 < \infty\}$. Characterize the families $\{a_i\}$ for which $f(E) \subset E$. Show that in this case $f|_E$ is continuous when E is given the topology of the norm $\|x\| = (\Sigma x_i^2)^{1/2}$.

(iii) For which families $\{a_i\}$ is $f|E: E \to E$ a compact mapping for this norm topology?

17.4 A compact embedding

Let C^1 denote the continuously differentiable maps $f: [0,1] \to \mathbb{R}$ with norm $\|f\|_1 = \|f\| + \|\partial f/\partial x\|$ and C^0 the space of continuous maps $f: [0,1] \to \mathbb{R}$ with sup norm $\|f\|$. Show that the inclusion map $i : C^1 \to C^0$ is compact.

Remark. This is a prototype of what is known as the Sobolev embedding theorems; see Yosida [1].

§18 THE OPEN MAPPING AND CLOSED GRAPH THEOREMS

The two theorems mentioned in the title of this section are cornerstones in functional analysis. We shall prove these theorems in the case of separable metrizable complete TVS's. For a discussion of recent work on weakening the hypotheses see Tagdir Husain, [1] _The Open Mapping and Closed Graph Theorems in Topological Vector Spaces_.

Our proof of the open mapping theorem is different than the standard one, with more of a topological flavor. The reader should supplement this with the classical proof.

18.1 _THEOREM (OPEN MAPPING THEOREM)_. _Let_ E _and_ F _be metrizable complete topological vector spaces. Let_ $f: E \to F$ _be continuous, linear, and one to one. Then either_

(i) $f(E)$ _is of first category in_ F

_or_ (ii) $f(E) = F$ _and_ f _is a_ TVS _isomorphism_.

We shall prove this below only in the case E is separable. The reason is that the proof is shorter and more topological than the classical method. (For the classical proof see Bourbaki,

Husain or Yosida). Recall that (by 15.17) the term "metrizable" means equivalently "metrizable topology" or "metrizable uniformity."

Before giving the proof we deduce a number of important consequences.

18.2 COROLLARY. In theorem 18.1, the assumption that F is complete may be omitted.

Proof. Let $\hat{F}$ be the completion of F, so that $\hat{F}$ is a metrizable complete TVS. Suppose $f(E)$ is not of first category in F. Then $f(E)$ is not of first category in $\hat{F}$, since F is dense in $\hat{F}$. (If $f(E) = \bigcup X_n$ and X_n is nowhere dense in $\hat{F}$ then X_n is nowhere dense in F). Then we conclude by 18.1 that $f(E) = \hat{F}$ so that $\hat{F} = F$. □

Notice that in the next corollary the one to one assumption is dropped:

18.3 COROLLARY. Let E and F be metrizable TVS's with E complete. Let $f : E \to F$ be continuous linear. Then either (i) $f(E)$ is of first category in F or (ii) $f(E) = F$ and f is an open mapping.

Proof. Let E_o = ker f , a closed subspace of E. Then E/E_0 is metrizable and complete by 15.18. Let $\pi : E \to E/E_0$ denote the projection which is open as was observed in 15.7. There is a continuous linear one to one mapping $\Psi : E/E_0 \to F$ such that $f = \Psi \circ \pi$. Now $f(E) = \Psi(E/E_0)$ so by 18.2, if f(E) is not of the first category then Ψ is an isomorphism. In particular Ψ is open and $f(E) = \Psi(E/E_0) = F$. Hence f is open. □

18.4 COROLLARY. (i) *Let E,F be metrizable TVS's with E complete and F of second category in itself. Suppose $f : E \to F$ is continuous linear and is a bijection. Then F is complete and f is bicontinuous.*

(ii) *Let E,F be complete metrizable TVS's with $f : E \to F$ a continuous linear bijection. Then f is bicontinuous.*

This follows at once from 18.2. From this we obtain restrictions on the possible topologies for a metrizable complete TVS as follows:

18.5 COROLLARY. *Let E be a vector space and* $\mathcal{T}_1$,

$\mathfrak{T}_2$ _two topologies making_ E _into a complete metrizable_ TVS. _Then if_ $\mathfrak{T}_1 \subset \mathfrak{T}_2$ _we have_ $\mathfrak{T}_1 = \mathfrak{T}_2$.

Proof. Apply 18.4 to the injection
$i : (E,\mathfrak{T}_2) \to (E,\mathfrak{T}_1)$. □

In general algebraic supplements are not always topological ones. Conditions ensuring this are:

18.6 _COROLLARY_. _Let_ E _be a complete metrizable_ TVS _and_ E_1, E_2 _closed subspaces of_ E _which are algebraic supplements_. _Then_ E_1, E_2 _are topological supplements as well_.

Proof. Consider the continuous linear map
$\varphi : E_1 \times E_2 \to E$; $(x_1,x_2) \mapsto x_1 + x_2$. Since E_1 and E_2 are closed, they are complete. Hence $E_1 \times E_2$ is metrizable and complete. Since E_1 and E_2 are algebraic supplements, φ is a bijection. Hence by 18.4 it is bicontinuous. □

To prepare for the proof of 18.1 we need the following topological theorem.

18.7 _THEOREM_. _Let_ E _be a complete separable_

metric space, F *a metric space and* $f: E \to F$ *a continuous one to one mapping*. *Then* $f(E)$ *is a Borel subset of* F . (*All Borel subsets of* F *arise this way*.)

We shall not give the proof of this theorem here. It may be found in Kuratowski [1], p. 487. (The proof is lengthy but is not difficult.)

18.8 DEFINITION. Let E be a topological space and $X \subset E$. We say that X has the *Baire property* iff there exists an open set $\omega \subset E$ and sets A , $B \subset E$ of the first category such that:
$X = (\omega \cup A) \; B$.

Roughly, X has the Baire property if it is "almost open", or open up to sets of first category.

18.9 PROPOSITION. *Let* E *be a topological space*. *Then the set of subsets of* E *which have the Baire property is a* σ-*field*. *In particular, every Borel set has the Baire property*.

Proof. If $X = (\omega \cup A)\setminus B$ then

$$E\setminus X = (\Omega \cup A')\setminus B'$$

where $\Omega = E \setminus c\ell(\omega)$, $A' = B \cup [(E \setminus \omega) \cap c\ell(\omega)] = \text{bd}\ \omega \cup B$ and $B' = A \setminus B$.

Now A' and B' are of first category since $(E \setminus \omega) \cap c\ell(\omega) =$ (boundary of ω) is of first category (it is nowhere dense). Hence if X has the Baire property, so does $E \setminus X$. Secondly, we show that sets with the Baire property are closed under countable unions. If $U_n = (\omega_n \cup A_n) \setminus B_n$ for ω_n open and A_n, B_n of first category, then

$$C = \bigcup\{U_n : n = 1,2,\ldots\}$$

satisfies $$(\omega \cup A) \setminus B \subset C \subset \omega \cup A$$

where $\omega = \bigcup \omega_n$, $A = \bigcup A_n$, $B = \bigcup B_n$. Here, ω is open and A, B are of first category. It follows that C has the Baire property.

Since open sets have the Baire property the last statement holds. □

Using this machinery we can now give our proof of the open mapping theorem 18.1 in case E is separable.

<u>Proof of 18.1 for E separable</u>. Suppose $f(E)$ is not of the first category. It will suffice to

prove that if $U \subset E$ is open then $f(U)$ is open, since if $f(E)$ is open in F, it is absorbing and so by linearity we would have $f(E) = F$. Thus for each $x \in U$ we want to show $f(x) \in \text{int}(f(U))$. We may assume, by effecting a translation, that $x = 0$. Choose a circled open neighborhood V of 0 in E such that $V + V \subset U$ and hence $(V+V)/2 \subset U/2 \subset U$. We shall show that in F,

$$0 \in \text{int}[f(V) + f(V)]/2 .$$

Now $f(V)$ is not of the first category, as

$$f(E) = \bigcup \{nf(V) : n = 1,2,\ldots\},$$

since V is absorbing in E and hence $f(V)$ is absorbing in $f(E)$.

Now by 8.7, V, being open, is homeomorphic to a complete separable metric space so that by 18.7, 18.9, $f(V)$ has the Baire property. Therefore we can write $f(V) = (\omega \cup A) \setminus B$ for A, B of the first category and ω open and non-empty (if $\omega = \emptyset$, $f(V)$ would be of first category). Also, we may assume ω is circled, since $f(V)$ is.

We now claim that $(\omega + \omega)/2 \subset f(V) + f(V))/2$ which will prove the assertion.

More generally, let $Y_1 = (\omega_1 \cup A_1) \setminus B_1$ and $Y_2 = (\omega_2 \cup A_2) \setminus B_2$ for A_i, B_i of first category and $\omega_1, \omega_2 \neq \emptyset$, ω_1, ω_2 open. Then $(\omega_1 + \omega_2)/2 \subset (Y_1 + Y_2)/2$. For, let $x_1 \in \omega_1$, $x_2 \in \omega_2$, so $x = (x_1 + x_2)/2 \in (\omega_1 + \omega_2)/2$. Let φ denote the map symmetric about $x : \varphi(y) = 2x - y$. The image of Y_1 contains points in common with Y_2, since A_i, B_i are of first category. (If not, $\varphi(\omega_1)$ and ω_2 would be disjoint up to sets of first category which cannot happen as they have a point and hence an open set in common).

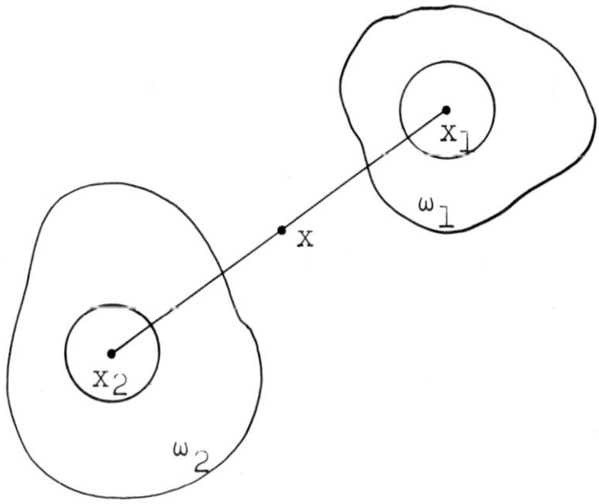

Figure 18.1

Suppose then that $\tilde{x}_2 \in Y_2 \cap \varphi(Y_1)$ and $\tilde{x}_2 = \varphi(\tilde{x}_1)$ for $\tilde{x}_1 \in Y_1$. Then $\frac{1}{2}(\tilde{x}_1 + \tilde{x}_2) = x$, so that $x \in (Y_1 + Y_2)/2$. □

For important classical applications of this theorem the reader should see Bourbaki and Yosida.

One of the most useful applications is to the closed graph theorem.

18.10 THEOREM (CLOSED GRAPH THEOREM). *Let* E *and* F *be complete metrizable* TVS's *and* $f : E \to F$ *a linear map*. *Then* f *is continuous iff* f *is a closed operator; that is the graph of* f ,

$$\Gamma_f = \{(x, f(x)) : x \in E\}$$

is closed in $E \times F$.

Proof. If f is continuous then Γ_f is closed (this is always true for F Hausdorff; exercise for the reader).

Conversely if Γ_f is closed then it is a metrizable and complete TVS. Consider the projection $\pi_1 : \Gamma_f \to E$; $(x, f(x)) \mapsto x$. Then π_1 is continuous, one to one and onto. By 18.1, π_1 is bi-continuous. That is, $f = \pi_2 \circ \pi_1^{-1}$ is continuous where π_2 is the projection on the second factor. □

Notice that "f closed" means that if $x_n \to x$ in E and $f(x_n) \to y$ then $f(x) = y$.

A typical application of the closed graph theorem is the following:

18.11 EXAMPLE. Let H be a Hilbert space and $T: H \to H$ a symmetric linear map (that is, for all $x,y \in H$, we have $\langle Tx,y\rangle = \langle x,Ty\rangle$). Then T is continuous.

Proof. Suppose $x_n \to x$ in H and $T(x_n) \to y$. Then for any $z \in H$ we have $\langle Tx,z\rangle = \lim_{n\to\infty} \langle x_n,Tz\rangle = \lim_{n\to\infty} \langle Tx_n,z\rangle = \langle y,z\rangle$. Hence $Tx = y$, or T is closed, so 18.10 applies. □

To convince oneself of the power of the closed graph theorem, one might try a direct proof in this example.

PROBLEMS FOR §18

18.1 An inequality of L. Hormander

Let E,F, and G be Banach spaces and $T: E \to F$ and $S: E \to G$ be continuous linear maps. Then there exists a constant C such that

$$\|T(x)\| \leq C(\|S(x)\|^2 + \|x\|^2)^{1/2}$$

and a constant D such that

$$\|S(x)\| \leq D(\|T(x)\|^2 + \|x\|^2)^{1/2} .$$

18.2 <u>Almost Positive Functions</u>

Let Ω be an open subset of $\mathbf{R}^n$ and $\mathcal{B}^2(\Omega)$ the Banach space of maps $f: \Omega \to \mathbf{R}$ which are of class C^2, for which the norm

$$\|f\|_2 = \|f\| + \|Df\| + \|D^2 f\|$$

is finite, where $\|\cdot\|$ is the uniform norm and D denotes the derivative. Also, let $B(\Omega)$ denote the space of bounded continuous functions $f: \Omega \to \mathbf{R}$ with the uniform norm (sup norm).

A linear mapping $A: \mathcal{B}^2(\Omega) \to B(\Omega)$ is called <u>almost positive</u> iff for all $f \in \mathcal{B}^2(\Omega)$, and each $x \in \Omega$, we have: $(f \geq 0$ and $f(x) = 0)$ implies $((Af)(x) \geq 0)$.

Show that if A is almost positive then A is continuous.

(An example of such a mapping is the Laplacian operator.)

§19 LOCALLY CONVEX SPACES

It is a general philosophical principle among mathematicians that in order to obtain interesting or deep theorems on topological vector spaces, one should assume that the space is locally convex. Indeed, for the Hahn-Banach theorem (see §21) this assumption is essential.

The principal result of this section is that a locally convex space (a TVS which has a base of convex neighborhoods at 0) may be described in terms of a family of semi-norms.

We shall begin our study with some of the elementary properties of convex sets and functions.

19.1 <u>DEFINITION</u>. Let E be a vector space (over $\mathbb{K}$, the real or complex numbers). A subset $X \subset E$ is called <u>convex</u> iff $(x,y \in X) \Rightarrow (\lambda x + (1-\lambda)y \in X$ for all $0 \leq \lambda \leq 1)$.

For a subset $X \subset E$, the <u>convex envelope</u> (or <u>convex hull</u>) of X is defined by

$$c(X) = \bigcap\{Y: X \subset Y \text{ and } Y \text{ is convex}\}$$

Recall that a subset $X \subset E$ is called a <u>cone</u> iff $\lambda X \subset X$ for all $\lambda > 0$. We say X is a pointed cone iff X is a cone and $0 \in X$.

A set $X \subset E$ is symmetric iff $X = -X$.

The following are immediate consequences, but are worthwhile observations (some were made earlier):

19.2 PROPOSITION. (i) The intersection of a family of convex sets is convex; in particular, $c(X)$ is convex

(ii) $c(X \cup -X)$ is the smallest symmetric convex set containing X

(iii) let X be a cone. Then X is convex iff $X + X = X$ iff $X + X \subset X$

(iv) let X be a convex cone. Then the linear space generated by X is $X - X$

(v) if X and Y are cones [resp. convex] then $X + Y$ is a cone [resp. convex]

(vi) let X be a convex cone. Then X is a vector space iff X is symmetric

(vii) let X be a convex pointed cone. The largest vector subspace of X is $X \cap (-X)$

(viii) let $\{X_i\}$ be a family of convex cones. Then $c(\bigcup X_i) = \Sigma X_i$ (finite sums of elements in the

X_i), <u>and is a convex cone</u>.

The next proposition deals with the interiors and closures of convex sets and is surprisingly delicate.

19.3 <u>PROPOSITION</u>. <u>Let</u> E <u>be a</u> TVS <u>and</u> $X \subset E$ <u>a convex subset</u>. <u>Then</u>

(i) <u>if</u> $x \in \operatorname{int}(X)$, $y \in c\ell(X)$, <u>and</u> $0 < \lambda < 1$, $\lambda x + (1-\lambda)y \in \operatorname{int}(X)$

(ii) $\operatorname{int}(X)$ <u>and</u> $c\ell(X)$ <u>are convex</u>

(iii) <u>either</u> $\operatorname{int} X = \emptyset$ <u>or we have</u>: $c\ell(\operatorname{int}(X)) = c\ell(X)$ <u>and</u> $\operatorname{int}(c\ell(X)) = \operatorname{int}(X)$.

<u>Proof</u>. (i) By a translation, we may suppose that $\lambda x + (1-\lambda)y = 0$. Hence $y = \alpha x$ for $\alpha = -\lambda/(1-\lambda)$. Since the mapping $z \mapsto \alpha z$ is a homeomorphism, there exists $z \in \operatorname{int}(X)$ such that $\alpha z \in X$ (the image of every neighborhood of x contains y and hence intersects X). Let $\mu = -\alpha/(1-\alpha)$ so that $\mu z + (1-\mu)\alpha z = 0$ and $0 < \mu < 1$. Therefore $U = \{\mu v + (1-\mu)\alpha z\colon v \in \operatorname{int}(X)\}$ is an open neighborhood of 0 which lies in X since X is convex

(ii) that $\operatorname{int}(X)$ is convex follows at once from (i). That $c\ell(X)$ is convex was observed in 15.6.

(iii) Suppose $\text{int}(X) \neq \emptyset$. Now $c\ell(\text{int}(X)) \subset c\ell(X)$. Conversely, let $y \in \text{int}(X)$ and $x \in \text{int}(X)$. Then $(1-\lambda)y + \lambda x \in \text{int}(X)$ for all $0 < \lambda < 1$ by (i). Letting $\lambda \to 0$ we see that $y \in c\ell(\text{int}(X))$.

Secondly we must show that $\text{int}(X) \neq \emptyset$ implies $\text{int}(X) = \text{int}(c\ell(X))$. It will suffice to show that $0 \in \text{int}(c\ell(X))$ implies $0 \in \text{int}(X)$. Let V be a circled neighborhood of 0, $V \subset c\ell(X)$ and choose $y \in V \cap \text{int}(X)$ (such a y exists since $c\ell(X) = c\ell(\text{int}(X))$). Therefore, $-y \in V$ since V is circled. Hence by (i), the segment joining y and $-y$ lies in $\text{int}(X)$. In particular, $0 \in \text{int}(X)$. □

These results are special for convex sets. For example, they fail for cones. (In $\mathbb{R}^2$, consider the first quadrant plus the whole x-axis).

The convex hull of a set is described as follows (the proof is immediate):

19.4 PROPOSITION. *Let* E *be a vector space. Then*

(i) $X \subset E$ *is convex iff for any* $x_1, \ldots, x_n \in X$ *and* $\lambda_1, \ldots, \lambda_n \geq 0$ *with* $\Sigma \lambda_i = 1$ *we have*

$$\sum_{k=1}^{n} \lambda_k x_k \in X$$

(ii) <u>for any</u> $X \in E$, $c(X)$ <u>consists of all elements of</u> E <u>of the form</u> $\Sigma\{\lambda_k x_k : k = 1,2,\ldots,n$, <u>for</u> $x_k \in X$, $\lambda_k \geq 0$, <u>and</u> $\Sigma \lambda_k = 1\}$.

Another useful fact is:

19.5 <u>PROPOSITION</u>. <u>Let</u> E <u>be a Hausdorff</u> TVS <u>and</u> $K_1,\ldots,K_n$ <u>compact convex sets in</u> E. <u>Then</u> $c(\bigcup\{K_k\})$ <u>is compact</u> (<u>and hence closed</u>).

<u>Proof</u>. Let $A = \{\alpha = (\alpha_1,\ldots,\alpha_n) \in \mathbb{R}^n : \alpha_i \geq 0, \Sigma \alpha_i = 1\}$ and $\varphi : A \times \Pi K_i \to E$; $\varphi(\alpha_1,\ldots,\alpha_n, x_1,\ldots,x_n) = \Sigma\{\alpha_i x_i\}$. By 19.4 (ii), we have $(A \times \Pi K_i) = c(\bigcup K_k)$. But φ is continuous and $A \times \Pi K_i$ is compact. □

For example the convex hull of a finite set is compact (and closed).

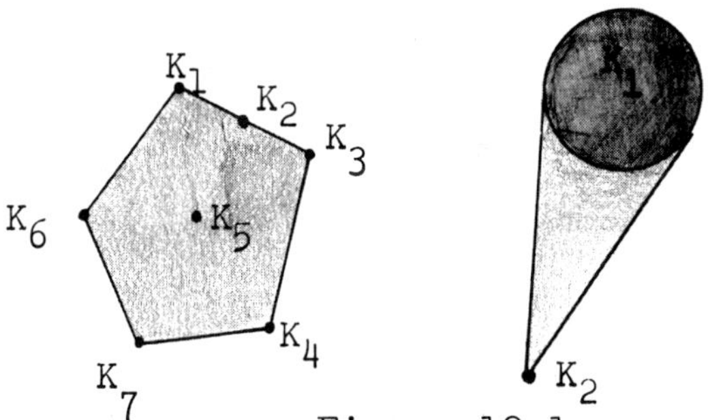

<u>Figure</u> 19.1

This proposition does __not__ mean that if K is compact then $c(K)$ is compact. In general, $c(K)$ is only pre-compact (problem 19.2). For example let X be the closed unit disc in $\mathbb{R}^2$ and $K \subset \mathcal{M}^1(X)$ the discrete measures supported by the boundary of X. By the theorem of approximation, $c(K)$ is dense in $\mathcal{M}^1(X)$, so cannot be closed.

19.6 __DEFINITION__. Let E be a vector space and $X \subset E$ a convex set. A map $f: X \to \mathbb{R}$ (or more generally with values in an ordered vector space) is called __convex__ iff $x,y \in X$, $0 \leq \lambda \leq 1$ implies $f(\lambda x + (1-\lambda)y) \leq \lambda f(x) + (1-\lambda) f(y)$. We call f __strictly convex__ iff $x,y \in X$, $0 < \lambda < 1$ implies $f(\lambda x + (1-\lambda)y) < \lambda f(x) + (1-\lambda) f(y)$.

A map $g: X \to \mathbb{R}$ is __concave__ [__resp. strictly concave__] __iff__ $-g$ __is convex__ [resp. __strictly convex__].

We let $C(X) = \{f: X \to \mathbb{R}: f \text{ is convex}\}$

Also, for a map $h: X \to \mathbb{R}$, let

$$A(h) = \{(x,y) \in X \times \mathbb{R}: f(x) \leq y\}$$

and

$$A'(h) = \{(x,y) \in X \times \mathbb{R}: f(x) < y\}$$

(the portion of $X \times \mathbb{R}$ above the graph of f; see diagram 19.2).

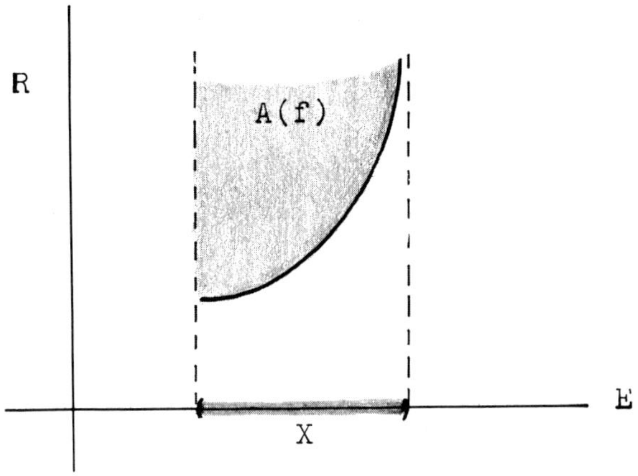

Figure 19.2

Notice that f is lower semi-continuous on X iff $A(f)$ is a closed set in $X \times \mathbb{R}$. [Proof: if $A(f)$ is closed then $\{x \in X: f(x) \leq \lambda\} \times \{\lambda\}$ $= A(f) \cap \{(x,y): y = \lambda\}$ which is closed. Therefore, $\{x \in X: f(x) \leq \lambda\}$ is closed and f is lower semi-continuous. Conversely, if f is lower semi-continuous and $(x_i, y_i) \to (x,y)$, $f(x_i) \leq y_i$ then $f(x) \leq y$ so $(x,y) \in A(f)$. Here we use the easy fact that f is lower semi-continuous iff $x_i \to x$ and $f(x_i) \leq \lambda$ implies $f(x) \leq \lambda$ (which is equivalent to: $\{x : f(x) \leq \lambda\}$ is closed).]

The characterization of convex functions is as follows:

19.7 PROPOSITION. *Let* E *be a vector space and* $X \subset E$ *be a convex subset. Let* $f : X \to \mathbb{R}$ *be a mapping. Then the following are equivalent:*

(i) f *is convex on* X

(ii) f *is convex on each line segment in* X

(iii) *for all* $x_1, \ldots, x_n \in X$ *and* $\alpha_1, \ldots, \alpha_n \in \mathbb{R}$ *with* $\alpha_k \geq 0$ *and* $\Sigma\, \alpha_k = 1$ *we have*

$$f(\Sigma\, \alpha_k x_k) \leq \Sigma\, \alpha_k f(x_k)$$

(iv) $A(f)$ *is convex in* $X \times \mathbb{R} \subset E \times \mathbb{R}$

(v) $A'(f)$ *is convex in* $X \times \mathbb{R} \subset E \times \mathbb{R}$.

In addition, each of these implies the following:

(vi) *for every* $\lambda \in \mathbb{R}$, $\{x \in X: f(x) \leq \lambda\}$ *is a convex subset of* X *and*

(vii) *for each* $\lambda \in \mathbb{R}$, $\{x \in X: f(x) < \lambda\}$ *is convex.*

This follows immediately from the definitions.

The set of convex functions is stable under many operations as follows:

19.8 PROPOSITION. *Let* E *be a vector space and* $X \subset E$ *a convex subset with* $C(X)$ *the set of convex real functions on* X . *Then*

(i) $C(X)$ *is a convex cone, that is*, $C(X)$ *is stable under addition and multiplication by* $\lambda > 0$.

(ii) *Suppose* f_i *is a net in* $C(X)$ *and for each* $x \in X$, $f_i(x) \to f(x) \in \mathbf{R}$. *Then* $f \in C(X)$.

(iii) *Let* $\{f_i\} \subset C(X)$, *and suppose* $f(x) = \sup\{f_i(x)\} < \infty$. *Then* $f \in C(X)$.

Proof. Each is obvious from 19.6 or 19.7. For example to prove (iii) notice that

$$A(f) = \bigcap \{A(f_i) : i \in I\}$$

which is convex. □

Of course for $g = \inf\{f_i\}$, $g \notin C(X)$ in general, as the union of convex sets $(A(g) = \bigcup A(f_o))$ is not in general convex.

Next we characterize continuity of convex functions:

19.9 *PROPOSITION*. *Let* E *be a* TVS *and* X *an open convex set with* $f: X \to \mathbf{R}$ *convex*. *Then the following are equivalent*

(i) f *is continuous*

(ii) f <u>is upper semi-continuous</u>

(iii) <u>there is a</u> (<u>non-empty</u>) <u>open subset of</u> X <u>on which</u> f <u>is bounded above</u>

(iv) <u>there is an</u> $x_0 \in X$ <u>at which</u> f <u>is continuous</u>.

<u>Proof</u>. It is clear that (i) implies (ii) and that (ii) implies (iii). Finally we must show (iii) implies (iv) and (iv) implies (i).

Suppose $x_0 \in X$ and f is bounded above by M, $M > 0$, on a neighborhood V of x_0. We may suppose $f(x_0) = 0$, $x_0 = 0$, and V is balanced (circled) because translations are homeomorphisms and preserve convexity. Let $0 < \varepsilon < 1$ and $x \in \varepsilon V$. Then by convexity, $-\varepsilon M \leq f(x) \leq \varepsilon M$ (Since $f(0) = f([x-x]/2) \leq [f(x) + f(-x)]/2$ so that $-f(x) \leq f(-x) \leq \varepsilon M)$, proving that f is continuous at x_0. Hence (iii) implies (iv).

Finally we show that (iv) implies (i). By the implication (iii) $\Longrightarrow$ (iv), it suffices to show that f is bounded above on some neighborhood of every point. Let f be bounded above on the neighborhood $x_0 + V$ of x_0. By a translation we may assume $x_0 = 0$. Let $y \in X$. Choose

$U \subset V/2$ such that $y + U \subset X$. Then if $z \in U + y, z = y + (1/2)z'$, $z' \in V$. Hence $f(z) \leq (1/2)f(z') + (1/2)f(2y)$ so f is bounded above on $y + U$. □

Notice that 19.9 holds in particular for linear functionals on E.

19.10 <u>COROLLARY</u>. <u>Let</u> $X \subset \mathbb{R}^n$ <u>be an open convex set and</u> $f: X \to \mathbb{R}$ <u>convex</u>. <u>Then</u> f <u>is continuous</u>.

<u>Proof</u>. There exists an open simplex in X; that is, (affinely independent) points $e_0, e_1, e_2, \ldots, e_n$ such that

$$V = \{x \in \mathbb{R}^n: x = \Sigma \lambda_k e_k; \Sigma \lambda_k < 1, \lambda_k \geq 0\}$$

is open and $V \subset X$. Since f is convex,

$$f(\Sigma \lambda_k e_k) \leq \Sigma \lambda_k f(e_k) < \Sigma|f(e_k)|$$

Hence f is bounded above on V and so is continuous by 19.9. □

In infinite dimensional spaces, 19.10 fails; in fact, there are non continuous linear maps, as the the reader well knows. Also, 19.10 fails (even in

$\mathbb{R}^n$) if X is not open (see problem 19.2).

Our next main goal is to show that the uniform structure of a locally convex TVS arises from a family of semi-norms, rather than a family of (translation invariant) pseudo-metrics.

13.11 DEFINITION. Let E and F be vector spaces over $\mathbb{R}$ and $X \subset E$ a pointed cone. A mapping $f: X \to F$ is called *positively homogeneous* iff $f(\lambda x) = \lambda f(x)$ for all $\lambda \geq 0$, $x \in E$.

If $X \subset E$ is a convex cone and $g: X \to \mathbb{R}$ is a map, we say g is *subadditive* iff $g(x+y) \leq g(x)+g(y)$, and we say g is *sublinear* iff g is positively homogeneous and subadditive.

19.12 PROPOSITION. *Let X be a pointed convex cone in E, a real vector space, and $f: X \to \mathbb{R}$ Then f is sublinear iff f is convex and positivel homogeneous.*

A map $p: E \to \mathbb{R}$ is a semi-norm iff p is sublinear and symmetric ($p(x) = p(-x)$). (See §2 for the definitions of norm and semi-norm).

Proof. Clearly f sublinear implies f is convex

and positively homogeneous. The converse holds by noting that

$$f(x+y) = 2f((x+y)/2) \leq 2\{f(x) + f(y)\}/2 = f(x) + f(y)$$

For the last statement we need only show that if p is sublinear and symmetric then $p(x) \geq 0$. However $0 = 0 \cdot p(0) = p(0) = p(x-x) \leq p(x) + p(-x) = 2p(x)$.□

Note that over a complex space we would require more to obtain $p(\lambda x) = |\lambda| p(x)$. Also recall that we allow generalized semi-norms to take the value $+\infty$. Of course sublinear functions need not be semi-norms (for example, any linear form).

Notice that if E is a vector space, $X \subset E$ a convex cone, and $f: X \to \mathbb{R}$ is sublinear, then for any $a > 0$, the sets

$$\{x \in X: f(x) \leq a\}, \{x \in X: f(x) < a\}$$

are convex and for every $x \in X$ there is an $\varepsilon > 0$ such that εx lies in the set.

19.13 <u>DEFINITION</u>. Let E be a vector space. For $x_1, x_2 \in E$, let $[x_1, x_2]$ denote the segment joining x_1 and x_2; that is, the set $\{\lambda x_1 + (1 - \lambda)x_2: 0 \leq \lambda \leq 1\}$.

A set $X \subset E$ is called <u>star-shaped</u> (at zero) iff there are $a_i \in E$ with $i \in I$ (for some index set I) and

$$X = \{[0,a_i];\ i \in I\} .$$

The <u>gauge</u> of X is defined as the map $p_X\colon E \to [0, \infty]$,

$$p_X(x) = \inf\{k > 0\colon x \in kX\} \leq +\infty .$$

For example the closed [resp. open] unit ball in $\mathbb{R}^n$ is star shaped. Its gauge coincides with the standard Euclidean norm. A similar phenomenon holds in general as follows:

19.14 <u>PROPOSITION</u>. <u>Let</u> E <u>be a vector space and</u> $X \subset E$.

(i) <u>The gauge of</u> X <u>is positively homogeneous</u>.

(ii) <u>every convex set containing</u> 0 <u>is star shaped</u>

(iii) <u>if</u> X <u>contains</u> 0 <u>and is convex then</u> p_X <u>is sublinear</u>

(iv) <u>if</u> X <u>contains</u> 0 <u>and is convex and symmetric then</u> p_X <u>is a generalized semi-norm</u>.

(v) <u>Assume</u> E <u>is a</u> TVS. <u>If</u> p_X <u>is convex</u>

<u>and the star shaped set</u> X <u>is closed and saturated</u>, <u>that is</u>, $X = \{x: p_X(x) \leq 1\}$, <u>then</u> X <u>is convex</u> <u>and</u> p_X <u>is lower semi-continuous</u>.

<u>Proof</u>. (i) For $\lambda > 0$, we have

$$\{\lambda k: k > 0, x \in kX\} = \{\ell > 0: \lambda x \in \ell X\}$$

so that, taking the inf, $\lambda p_X(x) = p_X(\lambda x)$.

(ii) is obvious, since in this case,

$$X = \bigcup \{[0, x]: x \in X\}.$$

To prove (iii) we must show that p_X is sub-additive. In fact, suppose $x \in \alpha X$ and $y \in \beta X$ so that $x + y \in \alpha X + \beta X = (\alpha + \beta)X$, for $\alpha, \beta > 0$, since X is convex. Hence $p_X(x + y) \leq \alpha + \beta$. Since this holds for all such $\alpha, \beta > 0$, $p_X(x+y) \leq p_X(x) + p_X(y)$ (to see that $\alpha X + \beta X = (\alpha + \beta) X$ note that $(\alpha + \beta)X \subset \alpha X + \beta X$ for any X and that $\alpha x_1 + \beta x_2 = (\alpha + \beta)y$ where $y = \alpha x_1/(\alpha+\beta) + \beta x_2/(\alpha+\beta) \in X$).

For (iv) note that for $x \in X$, $\{k > 0: x \in kX\} = \{k > 0: -x \in kX\}$ as X is symmetric.

Finally we prove (v). If $x_1 \in X$ and $x_2 \in X$ and $0 < \lambda < 1$ then $p_X(\lambda x_1 + (1-\lambda)x_2) \leq \lambda p_X(x_1) + (1-\lambda)p_X(x_2) \leq 1$ so $\lambda x_1 + (1-\lambda)x_2 \in X$.

If X is closed then $\{x \in E: p_X(x) \leq \lambda\} = \lambda X$, which is closed, so p_X is lower semi-continuous. □

The next proposition shows how a family of semi-norms defines a locally convex topology.

19.15 PROPOSITION. *Let* E *be a vector space and* $\{p_i\}$ *a saturated family f semi-norms which therefore generate a topology for* E (see 5.1). *Then* E *is a* TVS *and possesses a base of convex neighborhoods about* 0 (*and hence about each point*). *These neighborhoods are*:

$$X_{i,\varepsilon} = \{x: p_i(x) < \varepsilon\}$$

for all i *and* $\varepsilon > 0$. *These neighborhoods are also circled*.

Furthermore, E *is Hausdorff iff for each* $x \in E$, $x \neq 0$, *there is an* i *so that* $p_i(x) \neq 0$.

Proof. To show that E is a TVS, one can easily verify the conditions of 15.11.

Since each p_i is a semi-norm, and hence convex, the sets $X_{i,\varepsilon}$ are convex (this would be false if the p_i were merely translation invariant

pseudo-metrics). Indeed if $p_i(x_1) < \varepsilon$ and $p_i(x_2) < \varepsilon$ and $0 < \lambda < 1$ we have
$p_i(\lambda x_1 + (1-\lambda)x_2 \leq \lambda p_i(x_1) + (1-\lambda)p_i(x_2) < \lambda\varepsilon + (1-\lambda)\varepsilon = \varepsilon$.
The neighborhoods are circled since $p_i(\alpha x) = |\alpha| p_i(x)$ and are absorbing by positive homogeneity.

The last statement follows at once from 15.4. □

In fact the main goal of this section is to show that all locally convex spaces arise this way.

First we give a convenient criterion for continuity in terms of the semi-norms p_i.

19.16 PROPOSITION. *Let* E *and* F *be* TVS's *with (saturated) families of semi-norms* $\{p_i\}$ *and* $\{q_j\}$ *respectively, generating the topologies. Let* $f: E \to F$ *be a linear mapping. Then* f *is continuous iff for each* q_j *there is a* p_i *and a constant* M *such that*

$$q_j(f(x)) \leq M\, p_i(x)$$

for all $x \in E$.

Proof. The condition clearly implies that f is continuous at 0, and hence at every point. Conversely if f is continuous then for every q_j and

$\epsilon > 0$ there is a p_i and $\delta > 0$ such that $p_i(x) \leq \delta$ implies $q_j(f(x)) \leq \epsilon$. Let $M = \epsilon/\delta$ and we have for any $\lambda > 0$,

$$q_j(f(x)) = q_j\left(f\left(\frac{\delta x}{p_i(x)+\lambda}\cdot\frac{p_i(x)+\lambda}{\delta}\right)\right)$$

$$= \frac{(p_i(x)+\lambda)}{\delta}\, q_j\left(f\left(\frac{\delta x}{p_i(x)+\lambda}\right)\right)$$

$$\leq (p_i(x) + \lambda)\cdot \epsilon/\delta$$

which gives the result. □

19.17 <u>LEMMA</u>. <u>Let</u> E <u>be a</u> TVS <u>and</u> V <u>a convex circled neighborhood of</u> 0. <u>Then</u> p_V, <u>the gauge of</u> V <u>is continuous and finite</u>. <u>Also</u>,

(i) $V = \{x: p_V(x) < 1\}$ <u>if</u> V <u>is open</u>

(ii) $V = \{x: p_V(x) \leq 1\}$ <u>if</u> V <u>is closed</u>.

<u>Proof</u>. First, p_V is finite as V is absorbing. Secondly, p_V is convex by 19.14 and since it is bounded above on V, it is continuous by 19.9.

(i) Suppose V is open; let $x \in V$. Then there exists $\lambda > 1$ such that $\lambda x \in V$ by continuit of scalar multiplication. Thus $V \subset \{x: p_V(x) < 1\}$. Conversely, suppose that $p_V(x) < 1$. Then there exists $\lambda < 1$ such that $x \in \lambda V \subset V$ since V is circled.

(ii) Suppose V is closed. By definition, we have $V \subset \{x: p_V(x) \leq 1\}$. Suppose $x \notin V$. Since V is closed, there is a $\lambda < 1$ such that $\lambda x \notin V$ by continuity of scalar multiplication. This implies that $p_V(x) > 1$, and hence $V = \{x: p_V(x) \leq 1\}$. □

19.18 DEFINITION. Let E be a TVS. Then E is called a *locally convex* TVS, or briefly a LCS iff E has a base of convex neighborhoods of zero.

19.19 THEOREM. *Let E be an LCS. Then $\{p_V: V$ is a convex circled neighborhood of $0\}$ is a family of semi-norms generating the topology of E (in the sense of 5.1 or 5.4). Furthermore, the above set equals the set of all continuous semi-norms on E.*

Proof. The first part follows at once from the result that p_V is a semi-norm and for V open, $V = \{x: p_V(x) < 1\}$. Each p_V is certainly continuous as it is convex and continuous at 0 (observed in 19.17). Let q be a continuous semi-norm. Then $U = \{x: q(x) < 1\}$ is a convex circled neighborhood of 0 and has gauge q by 19.17. □

Next we consider conditions under which an LCS is metrizable or normable.

19.20 <u>THEOREM</u>. (i) *Let* E *be a Hausdorff* LCS. *Then* E *is metrizable iff the topology is generated by a countable family of semi-norms*.

(ii) *Let* E *be a Hausdorff* TVS. *Then* E *is normable iff* E *has a bounded convex neighborhoof of* 0 .

Proof. (i) is an immediate consequence of 15.17.

(ii) if E is normable, then E has a bounded convex neighborhood of zero, namely the unit ball.

Conversely, let U be a bounded convex neighborhood of 0 . Let $V \subset U$ be circled and W the convex hull of V . Then W is convex, circled and is bounded as $W \subset U$. Let p be the gauge of W . Since W is bounded, $\{n^{-1}W: n = 1,2,\ldots,\}$ is a neighborhood base of 0 and hence p generates the topology of E . Since E is Hausdorff, p is a norm and not merely a semi-norm. □

We have already seen several examples of

locally convex spaces such as $\mathcal{M}(E)$ for E locally compact Hausdorff, the C^k spaces (including $k = \infty$), Banach and Hilbert spaces.

19.21 DEFINITION. Let E be a Hausdorff LCS. We call E a Frechet space iff E is metrizable and complete.

For example, $\mathbb{R}^N$ and $\mathcal{C}^k(\Omega)$ $(\Omega \subset \mathbb{R}^n,\ k \leq \infty)$ are Frechet spaces. (The proofs are not hard). Note that $\mathbb{R}^N$ is not normable as every neighborhood of 0 is unbounded. One can similarly show that $\mathcal{C}^\infty(\Omega)$ is not normable.

Regarding complete spaces, we have:

19.22 THEOREM. (i) Let F_i be LCS's. Then ΠF_i is a LCS.

(ii) Let E be a Hausdorff LCS. Then E can be embedded in a product of normed spaces. In particular, the completion of E is a LCS.

Proof. (i) is immediate. For (ii) let p_i denote a generating family of semi-norms on E, and let R_i be the equivalence relation defined by xR_iy iff $p_i(x-y) = 0$. Then E/R_i is a normed space and by 5.19, E may be embedded in $\Pi E/R_i$. Hence the theorem. □

Note that E is metrizable iff E can be embedded in a countable product of normed spaces.

Regarding quotient spaces, note:

19.23 PROPOSITION. *Let* E *be a* LCS *and* $F \subset E$ *a subspace. Then* F *and* E/F *are* LCS's.

Proof. That F is a LCS is obvious, for if V is convex so is $F \cap V$. For the second part let V be a convex neighborhood of 0 in E, and $\pi : E \to E/F$ the projection. Then $\pi(V)$ is a convex neighborhood of 0 in E/F (the image of a convex set under a linear map is convex). □

On E/F the gauge associated with $\pi(V)$ is

$$p_{\pi(V)}(\tilde{x}) = \inf\{p_V(x): x \in \pi^{-1}(\tilde{x})\}$$

for $\tilde{x} \in E/F$. This coincides with the usual way of relating a norm on E to one on E/F.

Finally in this section we briefly consider barreled spaces and bornological spaces. In applications it is sometimes convenient to isolate these properties.

19.24 DEFINITION. Let E be a TVS. A barrel in E is an absorbing convex circled closed set. We say E is barreled iff each barrel is a neighborhood of 0 .

19.25 THEOREM. *Let E be a LCS which is a Baire space (or is of second category.) Then E is barreled. In particular Frechet spaces are barreled.*

Proof. Let D be a barrel in E . Then $E = \bigcup\{nD : n = 1,2,\ldots\}$. Since E is Baire, we have by 7.3, some $n_0 D$ has an interior point, say $n_0 y$. Then $y \in \text{int}(D)$. Since D is circled, $-y \in D$ and hence $0 \in \text{int}(D)$ as D is convex. □

For example $\mathbb{R}^I$ for any index set I is Baire by 24.12 and hence is barreled. The notion of barreled spaces is useful, for example, in proving that certain LCS's are reflexive. See problem 23.4 (Bourbaki calls a "barrel" a "tonneau").

19.26 DEFINITION. Let E be a LCS. We say E is bornological iff every circled convex subset which absorbs every bounded subset of E is a neighborhood of 0 .

There are three basic facts which we summarize in the following:

19.27 THEOREM. *Let* E *be a* LCS. *Then*

(i) E *is bornological iff every semi-norm on* E *which is bounded on every bounded set is continuous*

(ii) *if* E *is metrizable then* E *is bornological*

(iii) *if* E *and* F *are* LCS's *with* E *bornological and* $f: E \to F$ *is a linear map, then the following statements are equivalent*:

(a) f *is continuous*

(b) f *maps every bounded set into a bounded set*

(c) *for every sequence* $x_n \to 0$ *we have* $f(x_n) \to 0$.

Proof. (i) If p is such a semi-norm then $A = \{x \in E: p(x) \leq 1\}$ is a circled convex set (see 13.17) which absorbs every bounded set, for if B is bounded and $p(x) < M$ on B then $B \subset MA$. Hence A is a neighborhood of 0 , so p is continuous (see 19.9).

Conversely if A is circled and convex and absorbs bounded sets, then p_A is a semi-norm which is bounded on bounded sets. Hence A is a neighborhood of 0 as p_A is continuous.

(ii) Let U_n, $n = 1,2,\ldots$ be a decreasing neighborhood base of 0. Suppose A is convex circled and absorbs bounded sets. Then there is an n_0 so that $U_{n_0} \subset n_0 A$, which would prove the assertion; for if not, let $x_n \in U_n$ but $x_n \notin n\,A$ for each n. Then $\{x_n\}$ is bounded (countable sequence essential) but cannot be absorbed by A.

Finally we prove (iii). Clearly (a) implies (c). To show (c) implies (b), let $B \subset E$ be bounded and $x_n \in B$, $n = 1,2,\ldots$. If $\lambda_n \to 0$ then since B is bounded, $\lambda_n x_n \to 0$, so that $\lambda_n f(x_n) \to 0$. If $f(B)$ were not bounded there would be a neighborhood U of 0 such that $f(B)$ does not lie in nU for any n. If $y_n \in f(B)\ nU$ then $y_n/n \in U$ for all n. But by hypothesis, $y_n/n = f(x_n)/n$ converges to 0. Finally (b) implies (a), for if U is a convex balanced closed neighborhood of 0 in F, then

$q(x) = p_U(f(x))$ is a semi-norm on E. Now q is bounded on every bounded set B since $f(B)$ is bounded. Therefore, by (i), q is continuous. Therefore using 19.17 $\{x \in E: f(x) \in U\} = \{x \in E: q(x) \leq$ which is a neighborhood of 0. □

PROBLEMS for §19

19.1 The Finest Locally Convex Topology

Every vector space has a finest topology making it an LCS. Is it Hausdorff?

19.2 A Theorem of Mazur

(i) Show that the convex hull of a pre-compact set is pre-compact.

(ii) If U is open, is $c(U)$ open?

(iii) X circled $\Rightarrow c(X)$ circled.

19.3 Affine Functions on Convex Sets

Let X be a convex subset of a Hausdorff TVS E and $f: X \to \mathbb{R}$ an affine function. Show that: (i) if f is bounded below on some neighborhood of some point of X, it is also bounded below on some neighborhood of every point of X.

(ii) deduce that if X is compact and f

is of the first class (see 8.4), then f is bounded on X.

19.4 Bounded Affine Functions

Let X be a compact convex subset of a Hausdorff LCS E and let $f\colon X \to \mathbf{R}$ be affine. Show that: (i) if f is bounded below, it is bounded. (ii) this cannot be extended to convex functions on X, even $X \subset \mathbf{R}^2$.

19.5 Additivity of the Gauge

Let X be a convex cone in a Hausdorff TVS E and $f\colon X \to [0,\infty]$ a positive homogeneous map. Let $X_f = \{x \in X\colon f(x) \leq 1\}$. Show that X_f is a closed convex set in X with convex complement iff f is lower semi-continuous and additive ($f(x + y) = f(x) + f(y)$). Characterize the subsets of X which can be identified with some X_f.

19.6 The Box Topology

Let $\mathscr{J}$ be the topology on $\mathbf{R}^I$ (for some infinite set I) invariant by translation, and such that a set of neighborhoods of 0 is given by the sets $\prod_{i\in I} [-\epsilon_i, \epsilon_i]$ for $\epsilon_i > 0$. (This topology is called the box topology].

(i) Show that $\mathbf{R}^I$ is not a TVS under $\mathscr{S}$ but that it is a complete topological group.

(ii) Show that the restriction of $\mathscr{S}$ to the direct sum $\mathbf{R}^{(I)}$ makes $\mathbf{R}^{(I)}$ into a TVS and that $\mathbf{R}^{(I)} \subset \mathbf{R}^I$ is closed.

(iii) Let $\mathfrak{T}$ be the finest locally convex topology on $\mathbf{R}^{(I)}$ (defined by the set of <u>all</u> semi-norms - see problem 19.1). Show that $\mathbf{R}^{(I)}$ is complete under $\mathfrak{T}$, and that (when I is uncountable) $\mathfrak{T}$ is strictly finer than $\mathscr{S}$.

(iv) Verify that $\mathbf{R}^{(I)}$ is not a Baire space with the $\mathfrak{T}$ topology.

19.7 Analytic Functions

Let, in $\mathbf{C}$, $D = \{z \in \mathbf{C}: |z| < 3\}$ and

$\mathscr{H}(D) = \{f: D \to \mathbf{C}: f \text{ is analytic (holomorphic)}\}$.

Let $K = \{z \in \mathbf{C} : |z| \leq 1\}$ and put

$$p(f) = \sup \{|f(z)| : z \in K\}.$$

(i) Show that p is a norm on $\mathscr{H}$

(ii) The topology defined by p is not the same as the topology of uniform convergence on every compact.

(iii) Is $\mathscr{H}$ with the norm p complete?

INDEX

2-46 refers to page 46 of volume 2 etc.

APPENDIX

PROOF OF THE REMARKS FOLLOWING THEOREM 8.7

Let E be a _metric_ space which is strongly α-favorable for a player with perfect information (i.e. α's strategy can depend on previous moves). We want to prove that it is a G_δ in any metric space F containing it (hence homeomorphic to a complete metric space). This implies of course the same result for metric spaces which are strongly α-favorable for a player with no memory (which is equivalent to 'fortement tamisable'), that is, the result of theorem 8.7. It follows that the two notions of strongly α-favorable are in fact equivalent. We shall denote by $\delta(X)$ the diameter of any subset X of F.

From the fact that E is strongly α-favorable, it follows that there exists a family of open sets $\omega_{i_1 i_2 \cdots i_n}$ of F, where the sets of indices I and $I_{i_1 \cdots i_n}$ are defined by recursion with the following properties:

(1) $i_1 \in I$, and for any integer n,

$i_{n+1} \in I_{i_1 \cdots i_n}$

(2) $\omega_{i_1 \cdots i_n i_{n+1}} \subset \omega_{i_1 \cdots i_n}$;

$E \cap \left(\bigcup_{i_{n+1}} \omega_{i_1 \cdots i_n i_{n+1}} \right) = E \cap \omega_{i_1 \cdots i_n}$

and $\delta(\omega_{i_1 \cdots i_n}) < 2^{-n}$

and (3) for every infinite sequence $i_1, i_2, \cdots$ such that $i_1 \in I$ and $i_{n+1} \in I_{i_1 \cdots i_n}$ (called <u>compatible sequences</u>), the intersection of the sequence $(\omega_{i_1 i_2 \cdots i_n})$ is exactly a one-point subset of E.

Let us now, once and for all, endow each of the sets I, $I_{i_1 \cdots i_n}$ with a well-ordering. Then, for every finite compatible sequence $i_1 \cdots i_n$, we define by recursion the open set $A_{i_1 \cdots i_n}$:

$A_{i_1} = \bigcup_{j_1 < i_1} \omega_{j_1}$: and more generally

$$A_{i_1 \cdots i_n i_{n+1}} = A_{i_1 \cdots i_n} \cup \left(\bigcup_{j_{n+1} < i_{n+1}} \omega_{i_1 \cdots i_n j_{n+}}\right.$$

APPENDIX

Let $B_{i_1 \cdots i_n} = \omega_{i_1 \cdots i_n} \setminus A_{i_1 \cdots i_n}$. Finally, for every n, let X_n = (union of all $B_{i_1 \cdots i_n}$ relative to compatible sequences of length n) and let $E' = \bigcap X_n$. As $E \subset X_n$, for each n, we have $E \subset E'$. But, on the other hand, for every compatible sequence $i_1, \cdots, i_n$, the sets $B_{i_1 \cdots i_n j_{n+1}}$ are contained in $B_{i_1 \cdots i_n}$ and mutually disjoint; hence E' is the union of the sets:

$$x_{i_1 \cdots i_n} = B_{i_1} \cap B_{i_1 i_2} \cap \cdots \cap B_{i_1 i_2 \cdots i_n} \cap \cdots$$
$$\subset \omega_{i_1} \cap \omega_{i_1 i_2} \cap \cdots \cap \omega_{i_1 i_2 \cdots i_n} \cap \cdots$$

which is a one point subset of E. Hence $E' \subset E$, and finally $E = E'$.

If it can be shown that each X_n is a G_δ of F, the relation $E = \bigcap X_n$ will prove that E is indeed a G_δ of F.

That can be shown by recursion, using the following property.

LEMMA. Let X be a subset of a metric space F, and let (Ω_i) be a covering of X by open subsets of F; then the following holds:

(X is a G_δ of F) $\Longleftrightarrow$ (for all i, $X_i = X \cap \Omega_i$ is a G_δ of F) (in other words, the property of being a G_δ of a metric space is local).

Proof. It is an easy consequence of the fact that F is a topological space, each open subset of which is a paracompact, that we have: there exists an open covering (ω_j) of $\bigcup \Omega_i$ which is finer than (Ω_i) and locally finite. Each $(X \cap \omega_j)$ is a G_δ of F, hence is the intersection of a decreasing sequence $(\omega_{j,n})$ of open subsets of ω_j. For every n, let $O_n = \bigcup_j (\omega_{j,n})$. As each $x \in \bigcup \omega_j$ belongs only to a finite number of ω_j's, one has:

$$\bigcap_n O_n = \bigcup_j \Big(\bigcap_n \omega_{j,n} \Big) = \bigcup_j (X \cap \omega_j) = X$$

and hence X is a G_δ.

This completes the proof.

Note that the result 8.8 is now complete. More generally that argument shows every topological space F of the form F = f(E) with f continuous and open, and E strongly α-favorable for a player with perfect information, is of the same type. Hence any metric space F which is of that form f(E) is absolute G_δ: For instance, if E is complete, that gives a simple proof of the Hausdorff theorem; but this proof has the advantage that the Hausdorff theorem becomes a particular case of a more general theorem.

Another important application of the notion of strongly α-favorable is given in 27.9 (for a compact convex set X in an LCS, the extreme points $\mathcal{E}(X)$ is strongly α-favorable (for a player with perfect information). For a player with no memory, this is unknown (it is equivalent if $\mathcal{E}(X)$ is metrizable). Hence, if $\mathcal{E}(X)$ is metrizable, it is homeomorphic with a complete metric space. (If moreover it is separable, it implies that X itself is metrizable).